Confronting the Quantum Enigma

Confronting the Quantum Enigma

Albert, Niels, and John

David J. Kreiter

BINARY PULSE DISTRIBUTORS · IOWA CITY

Binary Pulse Distributors
Qreality1@aol.com

Printed in U.S.A.
ISBN: 978-0-615-48243-9

Contents

Quantum Enigma

THE BOUNDARIES between science and philosophy and between fantasy and reality have intrigued me throughout my life. As a child, I was fascinated by the vastness of the universe as I contemplated the possibility of the existence of life on the planet Mars, or on some planet in a distant galaxy, or the even more astounding possibility that extraterrestrials were making regular visits to our own planet Earth. My fantasies of space travel were realized as I witnessed the launch of the artificial moon, Sputnik, by the Soviet Union in 1957, and the subsequent launch of our own Explorer satellite the following year. I remember the first living creature—a dog—being propelled into orbit by the Soviets, followed by the chimpanzee, Ham, placed in orbit by the United States. I watched from a field on our farm in Iowa as Telstar, an inflatable communication satellite, traced across the starry night sky. Like many kids of my generation, I knew the names of every astronaut in the Mercury, Gemini, and Apollo programs, the names of the booster rockets that launched them and the names of the small capsules the astronauts were packed into like "spam in a can."

Slow paced cultural change typical of previous eras was replaced by a full-blown cultural revolution in the 60s and the 70s as innovations such as relay satellites and the transistor shrank our world. Music, fads, and fashion swept across the land in days or weeks instead of years. For the first time in history, we could all simultaneously share experiences and events such as ball games, the moon landing, rock-and-roll music, the Beatles and the British invasion, Woodstock, scandals, and protest movements. Real-time images of the Vietnam War compelled many of us to take action as we participated in anti-war demonstrations that sometimes escalated into battles with police and national guardsman in the streets and on college campuses.

This was also the time of the Cuban missile crisis. I clearly remember the fear I felt over a week that felt like months. I remember exactly where I was, on my sister's farm, when I experienced a strange feeling of resignation that I would be vaporized by an atomic bomb. I remember the press depictions of exploding atomic bombs, with the head of Albert Einstein and his famous equation wistfully floating above the mushroom cloud. I had a fearful fascination with this deadly, powerful creation that man had wrought upon the world, as well as with the man responsible for that simple mathematical equation: $E=MC^2$.

Those were exciting times, and the events of that period in history provided considerable distractions from my academic pursuits, but through it all, I never lost sight of my scientific and philosophical interests. Though I am still curious about the possibility of other life in the universe, my scientific interests have turned toward more fundamental questions of metaphysics—the study of reality.

It has been about twenty five years since I picked up a book written by Physicist Fred Alan Wolf titled, *Taking the Quantum Leap*. It was one of the first of its kind, written for the lay public, to succinctly lay out the philosophical implications of quantum physics. I was surprised to discover that I didn't need a degree in physics to understand the weird and fantastic ramifications of a new science that Max Planck almost apologetically discovered in the year 1900. I remember getting the same feeling of excitement I had experienced as a kid following the space program, when I first read and understood something called the two-hole or double slit experiment. The implications of this experiment for the nature of reality were so strange, and frankly unbelievable, that I wondered if what I was reading could possibly be real, verifiable science. I soon learned from subsequent reading that this was hard experimental physics. For the first time, I had come face to face with the quantum enigma.

I now understand why the father of the quantum revolution, Max Plank, a classical physicist, was dismayed by the implications of his own discovery; why Erwin Schrödinger said he was sorry he ever had anything to do with the whole quantum mess; why Albert Einstein said that if quantum theory meant that an electron "chooses" the time to move then he would rather be a cobbler or an employee in a gaming house than a physicist; and why Niels Bohr, the father of the atom, said that anyone who is not shocked by quantum theory has not understood it. Quantum physics is a complete and total departure from the universe we thought we knew.

Planck was dismayed because he had to abandon continuity, objectivity, and cause and effect as laws that govern our universe, and, though he launched the quantum revolution, he never came to accept fully anything but a classical description of his equation $E=h\nu$. Schrödinger was sorry, because the hope he had given classical physicists that a mechanistic description of the atom could be salvaged when he invented *wave mechanics* was dashed when he was forced to accept that the electron was not a "real" wave, but only a probability wave as physicist Max Born had demonstrated. Einstein was distraught at abandoning causality, and Bohr was shocked, because he understood the deep philosophical implications of an observer created universe.

These renowned scientists of the quantum revolution walked up to the edge of the chasm that separates the comforting logical world of classical physics and the disturbing illogical world of quantum theory, realized the inevitable conclusions that must be drawn about the nature of reality, and with great reluctance jumped into the frightening abyss.

Of the four scientists, only Bohr came to fully embrace the theory and the undeniable experimental evidence that supported it. He saw that the atom, which exhibited complementary attributes of wave and particle were part of the whole reality, and that any observation to detect either one of these properties could not be achieved without disturbing or changing the whole. He acknowledged that the role of the physicist was not to reveal how nature "is" but to describe reality based on the measurements we perform.

So what exactly *is* the *quantum enigma*? I define the quantum enigma as an attempt to reconcile the dichotomy between the scientific data gathered during the quantum revolution and our experience of the natural world. These scientific findings suggest that: 1) The universe has no independent existence, but has come into being as a result of observation and interpretation by observers; 2) The wave nature of matter and energy is not a "real" physical wave, but rather a probability wave; 3) The universe is acausal; that is, the ultimate reason for change and movement is chance and probability, not cause and effect; and 4) The universe is holistic and non-local: the discreteness of matter and energy which we intuit is an illusion.

Although the realm of the quanta surely blew me away, to be fair, classical physics is interesting in its own right. It gave us a new view of our place in the universe as a result of the Copernican revolution. Kepler's law of planetary motion demonstrated that the planets did not move in perfect circles. Galileo's experiments overturned centuries of belief about the effects of gravity on objects; he was one of the first to discover that motion was relative, and his optical telescope opened the heavens to exploration. Classical physics was responsible for Newton's three laws of motion and Einstein's counter-intuitive notions of space, time, motion, and gravity. Classical physics' mathematical predictions enabled mankind to launch men and machine into orbit, space probes to the outer reaches of our solar system, and even to land human beings on the moon. Classical physics was a triumph of the scientific process built over centuries, one step at a time like a giant pyramid, but the advent of quantum physics shook the very ground on which the pyramid stood. John Wheeler the renowned physicist, who procured Albert Einstein's chair at Princeton University and coined the term "black hole" said, "Nothing is more important about quantum physics than this: it has destroyed the concept of a world as 'sitting out there.'"

The atom, the foundation of material reality, was not a hard objective bit of matter with defined attributes located in some discrete location in space as traditional physics might have portrayed, nor were the actions of these tiny particles determinate (having a cause). The new physics revealed a more subjective ephemeral entity, whose attributes and position are indeterminate and subjective. That is, our own consciousness appears to have something to do with how material reality manifests. It now seems that quantum theory has put humans and all conscious beings back into the center of a universe from which Copernicus's heliocentric theory had banished us.

The very idea that underlying what seems to be a classical world of objectivity and causality, is a world of ghost-like, observer-dependent entities, was abhorrent for classical physicists such as Einstein, who proclaimed that he liked to think that the moon was there even when he was not looking. While Niels Bohr was not willing to suggest that macro objects such as the moon were observer-created, the Copenhagen interpretation of reality, which he and his colleagues formulated, confidently stated that *no subatomic event is a real event until it is an observed event.* This was a difficult proposition to accept for the mechanistic determinists who believed in an objective reality independent of observation.

In the early days, quantum theory had little experimental evidence to support it. Many of the ideas were nothing more than untested hypothesis, and as a result, its supporters found themselves under constant attack from Einstein and his like-minded determinists, who did not want to relinquish their grip on objective reality.

Einstein, the anointed leader of the determinists, along with his colleagues, formulated several thought experiments which threatened to overturn many of the new assumptions. One of the more notable arguments called EPR (Einstein, Podolsky, and Rosen), was a clever thought experiment portending to show that particles have objective attributes, not observer-created, dynamic attributes as the Copenhagen interpretation of reality suggested. The assault was so significant it left Bohr and his colleagues reeling, and threatened quantum theory in its entirety. Disagreements between the two camps lasted for decades and in something reminiscent of Custer's last stand, John Stewart Bell, while on sabbatical from CERN, came riding in to save the day for Albert Einstein, common sense, objective reality, and cause and effect. In what has been considered the most profound addition to quantum theory since its inception, Bell proved the opposite of what he had intended. Rather than confirming objective reality, *Bell's theorem of inequality* was a refutation of Einstein's classical notions and local causal connections. This revelation meant that connections between particles, while not technically violating relativity, were instantaneous and not dependent on local interactions.

Just a few years before Bell's death in 1990, experiments verified his mathematical theorem of inequality putting the final nail in the coffin of classical physics as a description of deep reality. It became clear from that time on that quantum physics was a complete theory and it appeared there was no turning back. The EPR thought experiment and Bell's theorem of inequality, with all of its implications for the nature of reality, will be thoroughly explored later when we meet the characters in this book: Albert, Niels, and John.

Philosophy for the Ages

ONE OF THE MOST important philosophical debates in recorded history commenced at the Solvay conferences in Brussels, Belgium in 1927 and 1930. The Solvay conferences, hosted by the eccentric rich industrialist Ernest Solvay, provided a stage for the most brilliant minds of the twentieth century. Four previous conferences dealt mostly with the fundamentals of quantum mechanics, but the 1927 and 1930 conferences were most notable because these meetings pertained to the *meaning* of the new quantum theory. Of the approximately 30 attendees to the conferences, over half were Nobel Laureates.

On one side of the great philosophical debate were the old guard determinists led by Albert Einstein, then 48 years of age and the youngest of his group. Lining up on his side were such luminaries as: Hendrik Lorentz, Max Planck, and Erwin Schrodinger to name a few. The determinists were products of the mechanistic, reductionist age of science and their philosophical beliefs reflected the scientific paradigm of that time in so much as: 1) Every action has a cause, 2) Everything can be understood as the sum of its parts, and 3) The observer is independent of that which is observed. Although Einstein and his colleagues all contributed to quantum mechanics, they believed that eventually hidden variables would be discovered that would save causality and objective reality. In a letter to his friend, Max Born Einstein wrote, "Quantum Mechanics is very impressive...but I am convinced that God does not play dice [with the universe]." Even more bothersome to Einstein was the idea of a participatory universe, a universe in which the observer determines reality. Einstein mused, "I cannot believe that a mouse can change the universe by merely looking at it."

On the other side of the debate were the young rogue indeterminists led by Niels Bohr and his supporters, including Werner Heisenberg, Max Born, Paul Dirac, Wolfgang Pauli, and Louis de Broglie. The indeterminists were converts to the new quantum revolution with all of its implications. They believed that chance and probability are the mechanism for change, and that there is no objective reality, independent of observation, at the subatomic level of reality.

Einstein was originally asked to give the opening remarks at the conference, but feeling unprepared he turned down the assignment. Instead, Bohr gave the opening presentation.

> *"He was unsparing in his description of what quantum mechanics had wrought. Certainty and strict causality did not exist in the subatomic realm, he said. There were no deterministic laws, only probabilities and chance. It made no sense to speak of a reality that was independent of our observations and measurements. Depending on the type of experiment chosen, light could be waves or particles."* (Isaacson, 345)

Einstein played coy at the 1927 conference, preferring to lie-in-wait for the chance to confront Bohr one-on-one. Each day during lunch, Einstein would challenge Bohr with thought experiments that seemed incongruent with quantum theory. With each new assault, Bohr and his indeterminist buddies would huddle to discuss the challenge and by dinner time he would have enough ammunition to strike down Einstein's opposition. Bohr had tremendous respect for the intellectual power of his friend Albert Einstein, and while Bohr's friends were not overly concerned that Einstein would overturn quantum theory, the insecure Bohr was not so certain (Isaacson, 346). Einstein, taking great pleasure in his sparring match with Bohr, was relentless. Each time the determinists thought they had knocked Einstein out, he would get up off the canvas and offer up an even more impressive challenge; and each time Bohr reeled from the punches.

> *One of the attendees to the conference said: Bohr was shaken by the challenge. "He walked from one person to another, trying to persuade them all*

that this could not be true, that it would mean the end of physics if Einstein was right," a participant recorded. "But he could think of no refutation. I will never forget the sight of the two opponents leaving the university club. Einstein, a majestic figure, walking calmly with a faint ironic smile, and Bohr trotting along by his side, extremely upset." (Isaacson, 348)

Bohr should not have been so concerned because the philosophical disputes served only to make quantum theory stronger, and though he was a poor public speaker and gave less than stellar presentations, most observers felt that Bohr and the indeterminists won the day at the 1927 conference. Einstein, however, was not ready to declare defeat. His next attack would come in the form of his 1935 EPR paper, which would confront two of the most bizarre predictions of quantum theory: *Indeterminacy and Non-locality.*

Heisenberg's indeterminacy principle suggests that the position and momentum of a particle cannot be known simultaneously, and the non-locality principle predicts that the measurement of one particle of an intertwined system will affect another particle from that system no matter how far apart they are separated in time and space. Einstein felt that a particle must have an objective position and an objective momentum even though it might be difficult to measure both at once, yet quantum theory states that the inability to know both position and momentum is not a result of inadequate technology, but a mathematical and philosophical reality. The more one knows about the position of a particle, the less one ***can*** know about the momentum and vice versa.

Even more bothersome to Einstein was the non-locality principle because he thought it violated relativity theory. Einstein assumed that by replacing Newton's idea of gravity which necessitated mysterious unmitigated forces acting instantaneously over vast distances of empty space, with an objective geometric explanation of gravity, he had once and for all gotten rid of what he called *spooky action at a distance.*

While at Princeton, he began to think seriously about these problems and in May of 1935 with the help of two colleagues, Nathan Rosen and Boris Podolsky, he published a paper called EPR, reflecting the initials of the three authors. The paper's title asked the question: "Can the Quantum-Mechanical Description of Physical Reality Be Regarded as Complete?" Specifically, The EPR thought experiment involved indeterminacy or as it is often referred: *the quantum measurement problem.* The EPR thought experiment is a bit cumbersome to understand, so years later David Bohm reformulated Einstein's thought experiment using the attribute of polarization, which is conceptually easier to understand. So we will also use Bohm's version of the EPR experiment for simplification. In his book *Quantum Reality,* Physicist Nick Herbert does an exemplary job of describing Bohm's set-up of the experiment. I will use his work as a guide in my description that follows.

To understand Bohm's version of the EPR experiment, we must first define the dynamic attribute called polarization. Polarization is the simplest attribute because it can take only two possible values for each direction measured: Either it is entirely polarized in the direction measured or it is entirely polarized at right angles to the direction of measurement (136). A critical thing to remember is that a photon has no polarization until it is measured. Polarization is a *dynamic,* or observer created attribute. The choice of the measurement determines the photon's angle of measurement. All one can learn when measuring a photon's angle of polarization is that it is either polarized in the direction one chooses to measure, or it is polarized perpendicular to the chosen measurement. If one decides to measure a photon's polarization at an angle of zero degrees, then the photon will either be polarized at zero degrees or it will be polarized perpendicular to that measurement—ninety degrees (136). (*Figure 1*)

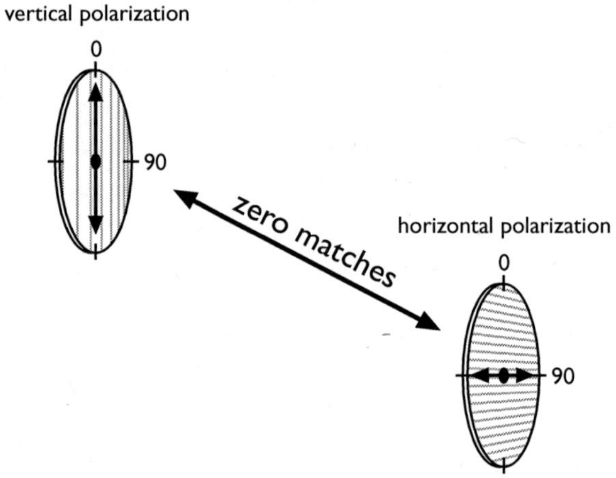

Figure 1.

The device used to actually measure polarization employs a calcite crystal. Calcite is a transparent mineral which bends light along one of two axes depending on the light's polarization. When a beam of light enters a calcite crystal, the light is split into two beams. Light polarized parallel to the optic axis of the crystal will not be bent and will travel straight through. Light polarized at right angles to the optic axis will be bent and take a deviant path. Calcite detectors normally employ a detection screen at the termination point to record which direction the photons took (202).

In Bohm's thought experiment, the light source will emit pairs of photons that are in a state of parallel polarization. Quantum theory and experiment confirm that twin photons in a state of parallel polarization have the same angle of polarization for all angles of measurement. The only thing that can be determined about the twin photons is that whatever angle of polarization one photon has after a measurement, the other photon will have the same angle of polarization when it is measured: This holds true no matter how far the photons fly from the source. Herbert points out that the polarization experiment has been demonstrated many times in the laboratory with the same results (203).

Bohm's experiment uses two detectors and a photon source that shoots photons in opposite directions. The experimenters at the detector site will decide which angle to measure the photons after they are already in flight. Since the detectors are not separated at great distances and because the speed of light is extremely fast, the measurement angle will have to be determined quickly. Detector A is several feet closer to the source than detector B. If the photon detector is set at a particular angle and the photon passes straight through the calcite crystal we know that the photon is polarized parallel with the detector setting, and it will go straight through to the "up" chamber. We shall designate this as an H for a "hit." If the photon's angle of polarization is perpendicular to the detector setting, then the photon will go into the "down" chamber and it shall be designated as an M for a "miss." The experimenters have a 50/50 chance of getting either a hit or a miss. Now suppose the two detectors are set to the same angle setting of 0 degrees and we shoot ten twin photon pairs from the source with each photon of the pair traveling in opposite directions toward their separate photon detectors. With both photon detectors at the same setting, we might get the following results by chance:

Detector A >>>>>>> H—H—M—H—M—H—M—M—M—H
Detector B >>>>>>> H—H—M—H—M—H—M—M—M—H

Note that the two detectors get the same results because they are set at the same angle. Any particular hit or miss is simply a result of a 50/50 chance. But we know that whatever angle of polarization is measured at detector A, its twin, arriving a fraction of a second later at detector B, will have the same angle of polarization. In the next part of the experiment, the photon detectors will be set perpendicular to each other so that detector A is set to 0 degrees and detector B is set to 90 degrees. The random results might look similar to the following:

Detector A >>>>>>>> M—M—M—H—H—M—H—M—H—M
Detector B >>>>>>>> H—H—H—M—M—H—M—H—M—H

Note that as expected, when the detectors are set perpendicular to each other, the results run counter to each other. A hit at the first detector will result in a miss at the second detector and vice versa. This stands to reason since the photons in a state of parallel polarization have identical angles of polarization when measured. Another essential point concerning polarization angles is that it is only the relative angle between the two polarizers that is important (218).

Now we can frame Einstein's argument, which in its essence is a refutation of the uncertainty principle. Because we used David Bohm's version of the EPR experiment, we have to change the rhetoric slightly to fit the experiment, but philosophically they are identical arguments. Einstein would say that the photons must have their angle of polarization determined at the source. If the experimenter at detector A sets his detector at 0 degrees and gets a *hit,* then the experimenter at detector B knows that, even before the twin photon reaches her detector, she will also get a *hit* when she measures the twin photon at 0 degrees. The polarization of both particles is known by measuring the first photon. This situation is identical in theory to knowing both position and momentum at the same time. Einstein claimed that since the polarization of the second photon is known in advance, it must already have its polarization determined at the source, and therefore it cannot be observer-dependent. In no way can the measurement of the first photon signal its polarization to the second photon before it is measured because the two are racing apart at the speed of light; therefore, he reasoned, the photon must have its polarization determined from the start.

Bohr and Heisenberg agree with Einstein that no signal passes between the two particles, but Bohr contended that neither photon has a polarization until it is observed (measured). Quantum theory is complete, they say; therefore, we must accept that neither photon has any angle of polarization until it is measured.

Technically speaking, there are a few more nuances to the arguments above and we will address them more thoroughly when we meet Albert, Niels, and John, but for now, suffice it to say, that this in principle was Einstein's difficulty with the new quantum theory. The news of the EPR experiment hit Bohr hard. In his book, *Einstein,* Walter Isaacson writes about the impact the EPR experiment had on Bohr:

> *When the EPR paper reached Niels Bohr in Copenhagen, he realized that he had once again been cast in the role, which he played so well at the Solvay Conferences, of defending Quantum mechanics from yet another Einstein assault. "This onslaught came down on us as a bolt from the blue," a colleague of Bohr's reported. "Its effect on Bohr was remarkable." He had often reacted to such situations by wandering around and muttering, "Einstein... Einstein... Einstein!" This time he added some collaborative doggerel as well: "Podolsky, Opodolsky, Lopodosky, Siopodolsky..."*
> *"Everything else was abandoned," Bohr's colleague recalled, "We had to clear up such a misunderstanding at once." Even with such intensity, it took Bohr more than six weeks of fretting, writing, revising, dictating, and talking aloud before he finally sent off his response to EPR* (451).

Bohr thought Einstein was just plain stubborn, and perhaps he was correct, but the conflict between the two was deeply rooted in the ancient Greek philosophy of *rationalism* and *dualism.* Rationalism supposes that empirical data is unnecessary to understand the world, because it is possible to comprehend the nature of reality through pure reason, and dualism contends that there is a distinction between the observer and the observed. Bohr was able to break free of the shackles of the philosophy of dualism—Einstein was not.

Rationalism, Reductionism, and Dualism

Rationalism was the philosophy of the Greeks, and the deductive process was their method of doing science. Plato believed that the rational mind was endowed with priori knowledge or truths that could directly access the world of "forms." With the help of mathematics, which he said belonged to the external realm, it was possible to circumvent the senses to gain access to the objective world.

The deductive method of science was first formulated by the Greek philosopher Thales (624–546 B.C.) and developed more fully by Aristotle (384–322 B.C.), who said all special sciences must begin with "self-evident" propositions. Those of us who studied geometry in high school are familiar with these self-evident propositions in the form of axioms or postulates. In this process one or more general propositions are stated and a specific conclusion is drawn, such as: All men are mortal; Socrates is a man; therefore, Socrates is mortal. The problem with self-evident propositions or postulates is that they are often faulty, and the conclusion is usually already assumed; otherwise, there would be no reason to state the postulate in the first place. As a result, nothing new is gained by this logical progression. Despite the fallibilities of the deductive method of science, many of the Ancient Greek principles were resurrected in the nineteenth and twentieth centuries by using the inductive method of science. Following are some examples of two distinct methods of observation, the deductive method and the inductive method, yielding very similar results:

1) Zeno of Elea: proposed through several paradoxical thought experiments that motion was not smooth and flowing, but discontinuous. In one such paradox he said:

> *"The arrow cannot fly. It cannot fly because an object that is behaving itself in a uniform manner is either continually moving or continually at rest. The arrow is certainly behaving itself in such a uniform manner. Now watch the arrow as it travels along its flight path. Clearly, at any instant, the arrow is occupying a given place. Therefore, if it is occupying a place, it must be at rest there. The arrow must be at rest at the instant we picture, and since the instant we have chosen is any instant,*

the arrow cannot be moving at any instant. Thus the arrow is always at rest and cannot fly." (Wolf, 17)

Twenty five centuries later, Max Planck, Werner Heisenberg, and Albert Einstein resurrected Zeno's idea, and overturned centuries of belief that motion was continuous. Planck's mathematical formulae $E = hv$, which described the emissions of heat and light from a black body showed that the emissions were not continuous, but came in "quanta" or chunks. Heisenberg, in his theory of indeterminism, found that nature revealed itself in a binary and complementary manner. One could either observe the momentum of a particle or the location of a particle, but not both simultaneously. And Albert Einstein, in his Nobel prize-winning paper concerning the photoelectric effect, theorized that light was not a smooth wave, but was made of individual units of energy called quanta.

2) Leucippus and Democritus (460–362 B.C.): believed that matter is composed of small discrete particles or atoms.

Albert Einstein in his 1905 paper on Brownian motion said that the buffeting of tiny pollen grains seen under powerful telescopes was the result of molecular motion. This convinced most physicists of atomic theory.

3) Pythagoras (569–500 B.C.): said that the universe operates according to whole numbers and their ratios. He believed that numbers are objective and belong to the external realm. The Pythagoreans chant was "Everything is number."

4) Plato (427–374 B.C.): thought that the knowing mind was endowed with certain truths that, with the help of mathematics, could directly access the external world of forms. He believed that numbers were independent of the senses: beyond the subjectivity of sight, sound, taste, and touch. While the senses are subjective and changing, he said, mathematics could lead us to the eternal unchanging truths of the realm of "forms."

Niels Bohr found that the electron could only occupy whole-unit orbits around the nucleus of an atom. Bohr solved the puzzle of why an electron does not spiral into the nucleus as a classical description would imply. An electron has a "lowest possible energy" described by Planck's constant "h." The quantum revolution has replaced the Pythagorean chant of "all is number" with a somewhat similar chant of "all is probability." Quantum processes such as nuclear decay and the quantum jumps of electrons from one orbit to the next are individually governed by the mathematics of chance and probability.

Planck used a mathematical formula $E=hv$ to interpret the odd emissions of heat and light from a blackbody, setting off the quantum revolution, and Werner Heisenberg used matrix algebra to describe the complex nature of the microscopic world of the atom.

5) Aristotle: believed in teleology, the idea that the future exists in reality and could influence the past. He said that the oak tree coaxes the acorn into becoming a tree.

Albert Einstein's block-time model of the universe suggests that the past and the future are every bit as real as the present. His idea has given support to those who advocate the teleological view point and its applications, to explain everything from the apparent backwards-in-time behavior of particles in quantum experiments, to cosmology, evolution, embryo development, and the laws of complexity.

Despite the confirmation of many deductive Greek ideas, the truth is that further progress could not be made without the new science of analysis and verification. By the sixteenth century, the Greek influence of rationalism and the deductive method began to lose influence in the shadow of the new scientific, cultural, and philosophical Copernican revolution. This transformative period completely changed the individual's perceived place in the natural world. In the Ptolemaic period, the role of the human was central to the universe. All heavenly bodies, which were considered to be on the same plane as earth, influenced human activities. Plagues, pestilence, and natural disasters, occurred as a result of man's inattention to the creator. Mankind was integral to the whole.

The Copernican era envisaged a stark contrast. We humans were thrust from the center of the universe whereupon we became mere spectators in the grand design. According to the reductionist doctrine, and the law of cause and effect, everything including human actions could be predicted if the initial conditions were known. By the dawn of the seventeenth century and the *age of reason*, the new mechanist age had taken a firm hold. The practitioners of this era such as Francis Bacon and Rene Descartes viewed the universe as a giant machine put in motion at the dawn of creation. The new science of analysis and verification was so successful during the mechanistic age that science and philosophy became separate. From that time forward philosophy would be relegated to merely posing the questions for empirical science to answer. It was truly a period of great scientific advancement but the philosophy of reductionism came at a price. We had for the first time become insignificant specks in an astronomically large universe.

The *age of reason* was, in many ways, a rebellion against the naiveté of *realism*. Realism is an intuitive philosophy which states that things are as they appear: a one-to-one correspondence between perception and reality. Since the mid seventeenth and eighteenth centuries, idealist philosophers such as Locke, Hume, and Berkeley have given us an understanding that much of the reality we perceive with our senses are attributes which are inferred upon the object and are not inherent in the objects themselves. These ideas were not without precedence however. The ancient Greeks realized that our senses cannot be trusted to give an accurate assessment of reality, yet Aristotle, one of the advocates of this idea, fell victim to the enticing notion that things are as they appear. It was his intuition rather than experimental evidence that convinced him that a feather and a boulder would fall at different speeds in a vacuum. It was also his intuition that if an object was dropped from a ship moving through the water, it would fall straight down and not at an angle relative to the water. It took centuries until Galileo disproved both of those notions through observation and experiment.

Many examples can be given as to the fallibility of intuition. Our intuition tells us that when we see a "tree" in the distance, it is exactly how we experience it, but the tree we see "out there" is really only an interpretation by our central nervous system of reflected wave lengths of light—presumably from a distant object—striking our retinas. Our intuition tells us that we are sitting in our stationary chair, which is in a stationary house, which is sitting on a stationary world. We know, of course, that this is not the case. Our chair and our houses are hurtling through space at phenomenal relative speeds. The earth is turning on its axis, the earth is traveling around the sun, the entire solar system is moving and bobbing around our galaxy, and in turn, our galaxy is headed toward an even larger galaxy that will eventually collide and merge with our own Milky Way. The common sense philosophy of *realism* was doomed to failure. Common sense, as Albert Einstein once said, is the set of all prejudices attained by the age of eighteen. As more intuitive notions were proven false, the scientists of the age of reason looked to the new science of analysis and verification in the form of the inductive method of science.

Specifically, the inductive process starts with recognizing patterns in an observation, formulating a hypothesis, gathering more data and using experiments, and, finally, formulating a general theory. This was a much more objective way of doing science. When controlled experiments were introduced into science it assured that one's own prejudices and erroneous assumptions did not enter into the process. A good example of this new scientific method was conceived and designed by A.A. Michelson and E.W. Morley in 1887.

The experiment by these two researchers was designed to detect the elusive ether of space that had been predicted by the ancient Greeks and had remained a credible idea right up into the end of the nineteenth century. It was thought that if light was a wave, as Thomas Young had shown through experiments in 1820, then light, like other known classical waves such as water waves or air waves, must have a medium through which to move. The ether of space fulfilled such requirements, but if it existed, it would have to have very unusual properties. It would have to have an ethereal quality, so as not to impede the movement of heavenly bodies, and yet be viscous enough to account for the extremely fast propagation of light.

The idea for the experiment was fairly simple; however, the implementation was extremely difficult because of the precision needed to measure the speed of light over a very short distance. In addition the instrumentation had to be shielded from almost all external vibrations. The experiment entailed focusing a light source onto a half-silvered mirror which split the beam of light into two beams traveling perpendicular to each other: one beam moving in the same direction of the earth's movement around the sun, and the other beam moving orthogonal to the earth's motion. As the two light beams reach the end of the apparatus, which are of precisely the same length, they each strike a mirror, sending them back to an interferometer. If the two beams are slightly out of phase, then an interference pattern occurs, indicating that the beam traveling lateral to the movement of the earth has taken longer due to the interaction with the ether. In the Michelson/Morley experiment, no interference pattern developed, and the experiment failed to detect the ether. Although the experiment did not definitively prove that the ether did not exist, it did demonstrate the objectivity of the experimental method when it went counter to the assumptions that were held. As Albert Einstein demonstrated with his special theory of relativity, the ether of space is an unnecessary construct. No such medium is required for the propagation of light.

Another advantage to the experimental method is that a theory is always subject to future falsification if it isn't supported by the data. Critics of experimental science often point to this uncertainty as fallibility. Creationists, for example, are fond of saying that evolutionary theory is "only a theory." While theories are not etched in stone, they are usually on very firm ground. Theories, after all, are hypothesis that are verified by experiment. This makes them particularly credible, especially if they stand the test of time. That does not mean that evolutionary theory will not be amended. It is quite possible that, as quantum theory broadened the scope of classical physics, so evolutionary theory will be found to be a subset of a more encompassing idea.

The philosophy of rationalism and deduction went hand in hand, and though the age of reason put an end to the deductive method for the most part, rationalism still remained an influential and natural part of scientific inquiry, and no one in history better personified this process than Albert Einstein. His five acclaimed scientific papers published in 1905, dealing with atomic theory, special relativity, and the photo electric effect, for which he won the Nobel Prize, contained few accreditations, because much of his work was done through pure rational thought. We should not overlook that fact, however, that his work, with the possible exception of his paper on special relativity, was predicated on existing experimental data.

From an early age Einstein wondered what it would be like to travel alongside a beam of light. If Maxwell's equations for the propagation of light were correct, then a wave of light could not be stationary, even as seen by someone traveling at the same speed as the light beam. Something had to give, and that something was time and distance. Light, he reasoned, must move at a constant speed for all observers in uniform motion because no one's frame of reference is more special than anyone else's frame of reference. Whether you are moving toward the light or moving away from the light, you could just as easily consider yourself to be at rest. As Galileo noted, there is no difference between an object in motion and an object at rest. In fact, there is no such thing as being at rest. Motion is the natural state of all objects in the universe.

It is noteworthy, that velocity is a measurement of distance divided by time, such as miles per hour or meters per second. And time itself is defined in terms of motion or distance: whether it is one revolution of the earth on its axis; one day, the motion of the earth around the sun; one year, or the number of vibrations of a cesium atom; 9,192,631,770, one second. In 1983, the international standard for the length of a *meter* was officially defined as: *The length of the path traveled by light in a vacuum during a time interval of 1/299,792,458 of a second.* In this particular case the velocity of light, a constant, is used to define distance. One meter is defined as the distance light travels in approximately three nanoseconds: (D = R x T). So this equation says one meter (D) is equal to the velocity of light (R) x (3 nanosecond). But we see that the velocity of light (R x T) has already been established in terms of distance (D). So, written out the equation would be: One meter (D) [the distance light travels in three nanoseconds] = R x T [the distance light travels in three nanoseconds] a self-referential equation. From this alone we can see that the speed of light is the only constant in the equation; therefore, time and distance must be relative terms, each dependent upon the other to establish meaning.

Just ten years after Einstein published his theory of special relativity, in a truly a remarkable and unprecedented example of the power of intensive rational thought (and incidentally, a confirmation of Plato's notion that the mind can directly access the external realm) Einstein published his revolutionary work on general relativity. Einstein knew that special relativity had to be correct, but it didn't seem to encompass Newton's law of gravity for a couple of reasons: First, Newton's gravitational force acted instantaneously over a distance. This violated special relativity's idea that no information can travel faster than the speed of light. Second, according to Newton's inverse square law, the force of attraction between objects diminishes with the square of the distance between them, yet special relativity demonstrated that time and distance are relative concepts. Einstein spent years trying to reconcile these incongruities until, in what he called the happiest moment of his life, a realization struck him like a lightning bolt. In that moment, he realized that gravity and accelerated motion were indistinguishable.

Newton and Einstein had two different ways to explain gravitational force. Galileo had recognized through experiment that heavy objects rolling down inclined planes fall at the same rate of speed as lighter objects. Newton explained this in terms of his second law of motion: Force equals mass times acceleration. Newton said that if one simultaneously drops a one pound mass and a two pound mass they will both hit the ground at the same time because, although the force of gravity pulls twice as hard on the two pound mass as the one pound mass, it takes twice as much force to accelerate the two pound mass as the one pound mass. The two forces, gravity and acceleration in accordance with his equation F=ma, cancel each other out and this happy coincidence is responsible for the two masses falling at the same rate.

Einstein saw the equivalency of the inertial mass and the gravitational mass as more than a coincidence. If the inertial mass and the gravitational mass are the same, he reasoned, they must be two aspects of the same principle. He said that there is no force acting on the two objects; rather, they fall at the same speed due to the geometry of space/time near the Earth. He made the *principle of equivalence* the cornerstone of his theory of general relativity. The following examples demonstrate this principle:

A person standing in an elevator in which the cable has snapped does not feel the force of gravity as it plummets toward the Earth any more than a person in intergalactic space. Whether one is in the falling elevator or in a spaceship moving in constant motion in intergalactic space, the travelers will not feel the effects of gravity, and they will float freely inside their elevator or the cabin of the spaceship in a weightless state.

Similarly, if one is in intergalactic space in a spaceship that is accelerating at 32 ft. per second, each second, the laws of physics are indistinguishable from one's experience of gravity while standing on Earth. If one drops a ball aboard this acceleration spaceship, the ball will hit the floor at the same time and rate of speed as if one were standing on Earth. A thrown object will take identical curved paths, whether in the accelerating rocket ship or on Earth and a horizontal beam of light will hit the opposite wall of the cabin slightly lower just as if gravity had bent it down. If one stands on a scale, one will weigh the same whether on Earth or accelerating at 32 ft. per second in a spaceship. In short there is no way to distinguish acceleration from gravity.

Einstein realized that if gravity could be transformed away, then it must not be a real force. He said that gravity is merely the geometry of space/time. Accelerated motion, far from gravity, is indistinguishable from accelerated motion in the presence of gravity. The moon takes a curved path around the Earth, not because there is some force pulling on it, but because it is taking the shortest path in non-Euclidean space. Light red-shifts in the presence of gravity, just as it does when emitted from an object that is accelerating away from an observer, and extrapolating from this fact, relative time slows in the presence of gravity just as it does for someone accelerating through space relative to someone in a uniform state of motion. It all seemed to work logically, but still, Einstein had no real proof until the opportunity presented itself in 1919, when Sir Arthur Eddington set out to provide experimental evidence for General Relativity.

During a total eclipse of the sun, Eddington and his scientific expedition observed and measured starlight bending around the sun in a process called gravitational lensing just as Einstein had predicted. When told of this experimental proof of general relativity, Einstein seemed unimpressed, and said that he would have felt sorry for the poor Lord if the experiment would have proven otherwise. He had no doubt that his theory, produced completely of rational thought, was correct, but his feigned disinterest in the experimental method was only a front. In fact, years prior to the successful Eddington expedition an astronomer named Erwin Freundlich had read of Einstein's predictions and was prepared to launch an expedition to the Crimea to view a total eclipse of the sun forecast for August 21, 1914. When it looked as if the expedition would not get off the ground because of a lack of funding, Einstein offered to pay the remaining costs for the expedition. It turned out that Einstein's offer was not needed, because by then, war had broken out and the expedition members were captured and held prisoner by the Russian army. They were later released in a prisoner exchange program. Perhaps the failure to test his theory was a blessing, because Einstein had yet to fully develop his theory. His prediction of how much the light would be bent was only half of what the experiment would have discovered. He later refined his theory to account for a two-fold effect of the bending of the star's light. One half of the effect was due to the gravitational time dilation, and the other half was due to the longer path light must take to reach earth due to the curvature of space/time. Einstein's gesture to pay for the expedition underscores the fact that although he was a rationalist, he knew the value of experimentation.

Much of the debate that transpired in later years between Niels Bohr and Einstein could have been averted had he lived to see John Stewart Bell's inequality theorem and its subsequent experimental proof. Einstein aspired to the rationalism of the ancient Greeks, but there is little doubt that he would have considered the experimental verification of Bell's theorem of inequality as the final word on the validity of quantum theory. Still, when he proclaimed that "the only incomprehensible thing about the universe is that it is comprehensible," Einstein must have wondered how it is possible, as the plutonian philosophy expunged, that the knowing mind can directly access the external world.

Of the two rungs of Greek philosophy: rationalism and dualism, it is dualism that has become synonymous with western science. Dualism hinges on the idea that the world is made up of two substances: mind and matter. Rene Descartes' (1591–1650) cogito "I think therefore I am," brought forth secondary dualisms. In modern terms Descartes' line of logic could be framed by the following syllogism: All matter of substance in the universe is subject to the law of cause and effect. I have the freedom to think any thought I please. Since my thought is independent of the laws of substance, mind and matter must be distinct. So Descartes' credo really created three dualisms: 1) Mind vs. matter (consciousness and matter are distinct aspects of the world) 2) Indeterminism vs. determinism, (mind is indeterminate and matter determinate) and 3) Observer vs. observed ("I" with independent thought am separate from the universe at large). In essence these three dualisms provided the battle ground upon which Einstein and Bohr fought.

Certainly, the process of dualism and reductionism has been extremely fruitful as a scientific endeavor. Reductionism gave us an understanding of the building blocks of matter, an understanding of chemistry and the Periodic Table of the Elements, the basic structure of all living organisms, the cell, and DNA. Reductionism's success is undeniable, but there is a limit to its utility.

One of the shortcomings of dualisms is that they tend to spawn a never-ending cascade of more dualisms. Besides the three dualisms listed above, the dualism of mind vs. matter has given rise to monism vs. pluralism. Monism, in turn, has fractured into two sub-dualisms: psychical monism, which suggests that the one substance in the universe is mind, and physical monism, which contends that the one substance of the universe is matter. One soon sees the futility of dualism and reductionism brought forth in the mechanistic age of reason. The limits of the reductionist process are laid bare in regard to biological life: We marvel at the biological diversity of life on our planet, and we have arbitrarily divided it into categories of kingdom, phylum, class, family, order, and species. Few would argue about the utility of these divisions for understanding evolutionary processes, yet as reduction proceeds to the molecular and atomic scales of nature, the divisions between the categories disappear completely. Of the 100 or so elements known to man, only about 20 make up all biological life on this planet. Only five elements— carbon, oxygen, hydrogen, nitrogen, and phosphorous—make up the four categories of macromolecules—carbohydrates, lipids, nucleic acids, and proteins, which run the machinery of all life on the planet. It isn't the diversity of life that is so surprising; it is the similarity.

As the raging river of reductionism was rushing over the scientific and cultural landscape, the under-current of *dualism synthesis* was slowly and methodically swimming its way upstream against the tide. Isaac Newton comprehended that the force that caused an apple to fall to the ground was the same force that kept the heavenly bodies in motion. In his famous and remarkable publication in 1687, called the *Principia,* he synthesized the entire motion of the universe using only three axioms or laws. And in what was certainly one of the most important contributions of the nineteenth century to our understanding of the nature of reality, James Clerk Maxwell united electricity and magnetism into the singular *electromagnetic field.* Einstein, the greatest synthesizer of dualisms the world had ever seen, united space, time, and motion in his special relativity theory, the consequences of which formed the equivalency of mass and energy in his famous equation. Then, in 1915, in an impressive demonstration of the power of rationalism, he united gravity and acceleration. By the end of the nineteenth century, the view of the universe was moving away from the dualist notions of reductionism toward a more subjective and holistic synthesis of reality. Why, the frustrated Bohr wondered, was Einstein able to unite space and time, gravity and acceleration, mass and energy, and yet was unable to go one step further to unite the dualism of the observer and the observed? Perhaps the reason lies in what each man took from the philosophy they read as young boys.

The achievements of both Einstein and Bohr can be directly accredited to the influence of their philosophical and scientific readings as they were growing up. Einstein's knack for seeing unity in nature came as a result of his boyhood reading of science books by Aaron Bernstein, whose writings emphasized the relationship between seemingly diverse fields of science. Both Einstein and Bohr read the Idealist philosophers such as Locke, Kant, Hume, and Berkeley and each reaped the ideas that would propel them to fame. From Kant, Einstein understood that the reality we perceive is a consensus reality filtered through our senses. From Hume, Bohr drew on the notion that cause and effect is a tentative prospect at best. The philosophers of the seventeenth, and eighteenth centuries had a profound effect on both Bohr and Einstein, but their divergent paths can be interpreted as a reflection of the separate paths taken by two of the early ancient Greek philosophers: Parmenides of Elea, in southern Italy and Heraclites in Asia Minor.

Parmenides was an early advocate of the belief that change is an illusion, and that permanence is the only true reality, an idea that formed the basis of western science, materialism, and the invariant laws of an objective universe. These were the ideas that Einstein pursued throughout his life. Heraclitus, on the other hand, thought that change like a burning flame or the current in a river are the natural state of the universe—the ideas that were the seed of the great eastern religions of Buddhism and Taoism (Wade, 20). Niels Bohr, a student of Eastern philosophy, acknowledged fully that his revolutionary idea of the wave/particle complementarity was a direct result of the ancient Buddhist and Taoist philosophy that nature is neither dualistic nor monistic, but rather a unity of opposing, natural, self-organizing forces which the Chinese refer to as the *ti* and the *t'ien*. These forces express themselves in the interplay between symmetry and chaos, between form and energy, and between determinism and chance (20).

One of the key beliefs of Tibetan Buddhist practitioners is the idea that nothing has independent existence. This idea has found acceptance in modern science. The most fundamental particle known to science, the *quark* has no independent existence outside of the atom's nucleus. The quark, like the space/time, particle/wave, and the position/momentum complementarities, only has meaning within the construct of an undivided whole. They are the *ti* and the *t'ien*, the yin and the yang of Eastern philosophy that Bohr knew so well. Unfortunately, Einstein did not see his space/time synthesis as a complementarity. Instead, he saw the space/time continuum as an aspect of an objective universe independent of observation.

How will the philosophies of Albert and Niels change when they come together once again at Princeton University? Will their philosophies become even more entrenched, or will they merge? Will Niels remain frustrated with his friend Albert and pace back and forth muttering, Albert...Albert... Albert... or will they become even closer friends this time around? I can hardly wait to find out, but before we do, it is necessary to address the quantum experiments that are at the very heart of the debate that pitted Einstein and Bohr against each other. In the next chapter we will address the famous two-hole experiments. While the two-hole or double slit experiments, as they are often called, require very sophisticated equipment and are extremely difficult to carry out in the laboratory, they are conceptually easy to understand with a bit of contemplation. Niels Bohr once said that anyone who is not shocked by quantum theory has not understood it, and Richard Feynman said, he thought it safe to say that no one understands quantum mechanics. Feynman says of the two-hole experiments in his "lectures," this is "a phenomenon which is impossible, absolutely impossible, to explain in any classical way, and which has in it the heart of quantum mechanics. In reality, it contains the only mystery...the basic peculiarities of all quantum mechanics" (Gribbin, 164). Let's take a look at the two-hole experiments to get an idea of the strange world of the quanta.

Quantum Experiments

THE FOLLOWING single-hole and double-hole experiments are taken in part from descriptions by Nick Herbert in his book *Quantum Reality* and John Gribbin in his book *In Search of Schrodinger's Cat.* To help understand the two-hole or double-slit experiment, I will first describe an experiment called the one-hole experiment. Both experiments use similar apparatus, which will include an electron gun capable of firing a beam of electrons or individual electrons, a phosphorous screen which records the electron strikes, a metal barrier with one tiny hole through which electrons can pass, and a photographic plate.

The electron gun is similar to those in any cathode ray television. It is equipped with a control knob which, when turned on "high", allows the gun to fire a beam of electrons and, when turned on "low", will fire individual electrons. The phosphorous screen, which is much like the cathode-ray tube in an analog television set, glows when struck by electrons. Phosphor is a molecule which usually resides in a ground state, but when hit by an electron changes to a high-energy or "excited" state. The excited molecule releases energy in the form of a photon of light, and again returns to the ground state. The emission of photons by many phosphorous molecules is what causes the image on the screen. The barrier which will be placed between the gun and the target screen has a shutter-like hole whose size can be adjusted by a control. And finally, a photographic plate placed behind the screen will record electron strikes as they accumulate to produce a permanent record as an image on film (Herbert, 58).

The One-Hole Experiment

For the first part of the one-hole experiment, the barrier will be temporarily removed so that the electron gun is on one side of a room and the phosphorous screen is on the other side. If the electron gun is turned on so that it shoots a beam of electrons, we observe the effect as a bright dot on the screen. If the control on the gun is turned up so that more electrons are fired, the spot brightens, but the size of the spot remains the same. If the control on the gun is turned down so that just one electron at a time is fired, we can observe the individual electron strikes on the screen, and we can count them as they hit. This experiment demonstrates the particle-like nature of electrons (60). (*Figure 2*)

Figure 2. One-hole experiment

Now, let's put in place the metal barrier with only one hole, and open the hole to see what happens. When the gun is turned on to fire a stream of electrons, we observe the same pattern on the screen as before. If the control on the barrier is gradually turned down to decrease the size of the hole, the spot on the screen will slowly shrink in size, but then, when the hole reaches a certain diameter, the spot begins to grow larger, and finally, when the hole is closed even further, the spot disappears and we observe a series of bright and dark rings. These rings, called an Airy pattern, resembles an archery target and were mathematically calculated by George Airy in 1835, to demonstrate what happens to any classical wave, such as water waves or air waves, as it passes through a hole—bumping, colliding, and deflecting each other. This part of the experiment demonstrates that the electrons also have a wave-like nature (62).

So far, nothing too strange is happening as the electrons are behaving in a classical manner. But now let's change the experiment by turning down the control of the electron gun to fire one electron every minute, and press our photographic plate against the back of the screen so that each individual electron strike can be recorded on film. This should eliminate any chance that the crowded-together electron particles will interfere with each other as they did when a beam of electrons where fired. After several days the electron strikes accumulate on the plate. What we observe is mysterious. The individual electron strikes have produced the same Airy pattern as when a beam of electrons was fired. What did the electron interfere with when going through the hole? This experiment demonstrates that the wave-particle aspects of an electron coexist as it travels through the hole on its way to the screen, and when observed or measured the statistical wave collapses into a definitive particle (66).

The Two-Hole Experiment

With the two-hole or double slit experiment, the experimental arrangement will be the same as with the one-hole experiment; however, this time, a barrier with two holes will be utilized. Turning on the electron gun, we fire a beam toward the screen with both holes open on the barrier. As the electrons pass through the holes, the waves from each hole merge and interfere with one another causing a wave interference pattern on the screen. Now, if the control on the gun is turned down so that only one electron at a time is fired, we notice that after several thousand individual electrons have been fired the same wave interference pattern shows up on the photographic plate. This is surprising, because we would expect that a single electron would go through one hole or the other, not both. Why would we get an interference pattern? If this same experiment is repeated with just one of the two holes open, the Airy pattern returns. If the electron only goes through one hole, then why should it make any difference whether one or both holes are open? (Gribbin, 170) (*Figure 3*)

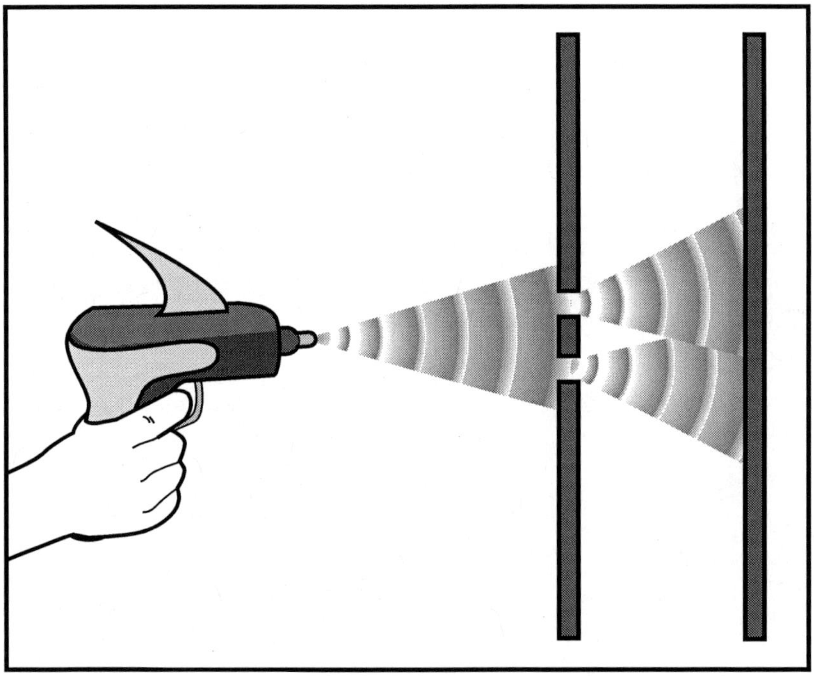

Figure 3. Two-hole experiment

The only conceivable explanation is that somehow each individual electron is going through both holes at once and interfering with itself. To determine if this is in fact what is happening, we can devise a simple experiment to observe the electrons as they pass through the holes. To achieve this test an observation experiment is set up on the far side of the barrier which actually counts the electrons as they pass through the holes. If a single electron is passing through both holes simultaneously, the event can be observed and recorded. But after firing one thousand single electrons with both holes open on the barrier, we find as statistically expected, that about five hundred electrons pass through each hole. Never does one electron pass through both holes simultaneously. Stranger still, when the electrons are observed and allowed to pass on to the screen, the interference pattern disappears, and there is an individual particle build-up on the screen! The observational counting experiment is set up after the electrons pass through the holes. How does the electron "know" in advance that it will be counted? Why should the observation affect the behavior of the electron before it gets to the holes when the observation takes place after the electrons pass through the holes (171)?

The results of this experiment are similar with any subatomic particle whether it is an electron, proton, neutron, or a photon of light. The subatomic particles behave one way when they are observed and another when they are not observed. And oddly enough, the observation can occur well after the particle should already have made a "choice" as to its route to the screen. "The electrons not only know whether or not both holes are open, they know whether or not we are watching them, and they adjust their behavior accordingly" (171).

The Photon Experiment

The following description of the photon experiment is taken, in part, from Roger Penrose's book, *The Emperor's New Mind.* This experiment is a variation of the famous two-hole or double slit experiment, and conceptually easier to understand. In this experiment photons of light are used rather than electrons. The setup for this experiment includes: a light source, which is capable of emitting a beam of light or a single photon of light, two fully-silvered mirrors that reflect all of the light shone on them, and a half-silvered mirror that statistically reflects one-half of the light shone on it, while allowing half of the light to pass through. And finally, a point where the photons meet and are allowed to interfere with one another. This interference pattern is then recorded on a light sensitive screen as in the originally conceived two-hole experiment. (*Figure 4*)

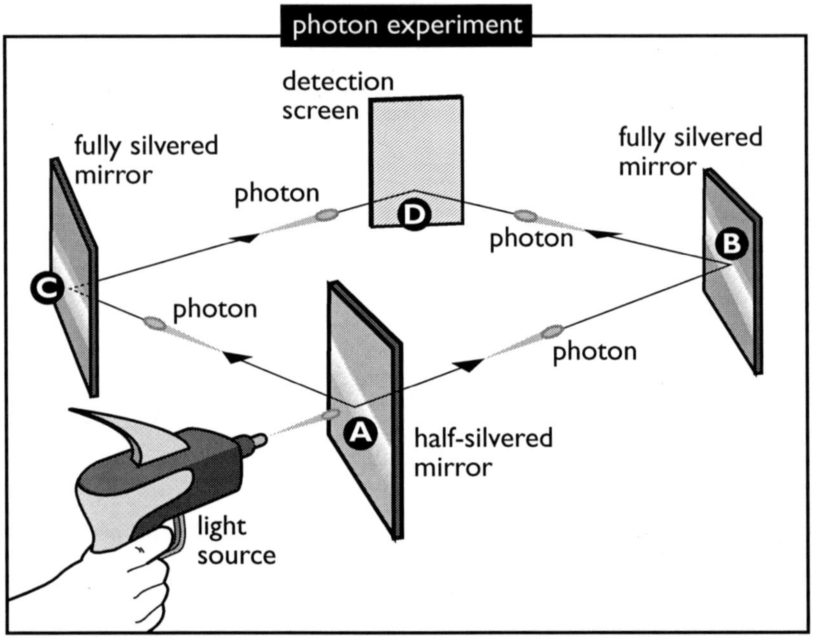

Figure 4. Photon experiment

If the light source is "turned on", a beam is emitted. As expected, one-half of the photons are reflected from the half-silvered mirror at point A up toward the fully-silvered mirror at point C, where they are reflected to the detection screen at point D. At the same time, one-half of the photons have passed through the half-silvered mirror at point A and continue to point B, where the photons are in turn reflected by the fully-silvered mirror to the screen at point D. Since the distances are identical for the two separate beams, the photons interfere with each other, and this interference pattern is recorded on the screen. So far all is as would be expected.

But now, if the volume control on the light source is turned down to allow only a single photon to pass, things become interesting. Will the photon pass through the half-silvered mirror or be reflected? If the pattern on the screen is examined, we discover a perplexing thing: An interference pattern has formed when a single photon is fired. It appears that the single photon must have taken both paths and interfered with itself.

In an attempt to make sense of the photon's behavior, an absorption screen is placed at point B, which will act like a detector. The photon will be absorbed and recorded if it takes path B. If the light source is again turned on so that just one photon of light is emitted every minute, it becomes clear that the photons take route B about 50 percent of the time, and route C about 50 percent of the time, as would be statistically expected. When the photon does not take path B, the screen at point D records the hit, which means the photon took route C. If the absorption screen is placed at point C, instead of point B, we get the same statistical results as before. If an absorption screen is placed at both points B and C, the photon is always absorbed and there is no interference pattern at the detection screen at point D.

It appears that when no measurement is made, the photon takes both paths simultaneously and interferes with itself at the screen. When a measurement at either point C or point B occurs, the photon "chooses" one path or the other. Even if the absorption screen is put into place at one of the two points C or B, at the very last instant after the photon is in flight, the same results occur. How does the photon "know" in advance if a measurement will be made (Penrose, 255)?

The interpretation of these particle experiments opens a chasm between the classical and quantum worlds. The temptation to see these experiments in terms of a teleological backwards-in-time scenario is forbidden by Relativity in that a cause cannot precede an effect. A classical explanation for these experiments is simply not possible, but a quantum explanation offered by Albert Einstein's good friend, physicist Max Born, can help us see through the veil. In one of the most remarkable contributions to quantum theory, he suggested that the electron wave is not a real wave but a statistical wave. His statistical explanation would mean that the electron in the "two-hole experiment" and the photon in "photon experiment" do not take both paths; their statistical wave takes both paths. An observation at one of the two routes collapses the statistical wave into a definitive particle. This idea raises several questions: First, what exactly is a statistical wave? Is it something physical? Is mathematics an objective aspect of the universe? Second, what constitutes an observation? Is it the measuring device? Is it consciousness? We just stated that a classical explanation of the experiments is impossible, so how can we say that a measurement or observation *causes* a change in the behavior of the subatomic particles? Many interpretations of the two-hole experiment still cling to the notion that it is the consciousness of the observer that "forces" a change in the behavior of the particle, yet Niels Bohr emphasized that our choice of measurement determines only which aspect of nature is revealed. Recent updated versions of the photon experiments leave doubt that an observer is required to collapse the statistical wave function:

In a "Wheeler-style," delayed-choice experiment conceived by Marlin Scully and performed at the University of California at Berkeley, polarized photons of light were used to verify the backwards-in-time effect in "erasing the past." In the following summary Paul Davies explains that the "observer" was entirely eliminated in one such experiment. We will revisit this experiment in more detail later.

> *Surprisingly, it is not necessary for the experimenter actually to go ahead and measure the photons' polarization to determine the paths they have taken for the change in detector behavior to be observed. The mere threat of obtaining such information is enough to destroy the ghostly superposition of hybrid phantom realities. It is our potential knowledge of the quantum system, not our actual knowledge, that helps decide the outcome.* (Davies, 171)

In the reference above, a delayed-choice experiment was set up in which there was no measurement or observation—only the possibility for extracting information about the path of the photon. Despite the absence of the observer, the photon behaved as if a measurement was taken. This experiment suggests that it is not necessarily the knowledge of the observer that is significant, but rather the "knowledge" of the universe at large. Terms such as "threat," "potential," and "knowledge" used to interpret this experiment are not normally used to describe a cold, objective, mechanistic universe. These words are usually reserved to define sentience. Still many were not convinced that nature could be so fickle.

In 1964 a little-known scientist from Belfast, Northern Ireland, took a Sabbatical from CERN, to take another look at Einstein's objections to the meaning of quantum theory that were laid out in his famous EPR paper. Like Einstein, John Stewart Bell was uncomfortable with the implications of quantum theory, and he wanted to see if it was possible to save objective reality. As it turned out, Bell's theorem proved the opposite of what he had hoped. His mathematical proof showed that nature is non-local and acausal.

In 1982 an experimental verification of Bell's theorem of inequality was performed at the University of Paris-South by Alain Aspect which put most doubts to rest. The experimental results demonstrated that a measurement performed on one of a pair of correlated photons instantaneously affects the second particle regardless of their separation in space/time. This was the strongest evidence yet that at the most basic level of reality the universe is an undivided whole.

By definition, objective scientific inquiry can be carried out only if there is a division between the observer and the observed. Science can claim knowledge of physical reality only when the predictions of a physical theory are validated by experiment. In a non-local universe, the indivisible whole cannot be measured or observed; therefore, nothing can be said about the character of this reality (Nadeau, 4). The tripod upon which reductionism rests has been whittled away: everything can be understood as the sum of its parts, everything has an initial cause, and the observer is independent of the observed. This is not just a blow to causality; it is a defeat for the very foundation of reductionism and objective reality.

The reality that science has exposed is far different from the one we intuit. It appears that the hard cold reality we perceive with our senses is nothing more than illusion as many ancient traditions have advocated. These assumptions have been given credence by researchers such as Tom Siegfried, who suspects that the universe is simply a vast information field:

> *Studies of the physics involved in information processing (especially with computers) have now solidified the conclusion that information is more than a metaphor. It is not something abstract and subjective, but as real as atoms, energy, or rocks. Information can be quantified. It is always embodied in some physical representation, whether ink on paper, holes in punch cards, magnetic patterns on floppy disks, or the arrangement of atoms in DNA. To me, this is one of the deepest and most important while least appreciated and least understood discoveries of modern science: Information is physical.* (Siegfried, 66)

Professor John Archibald Wheeler, founder of the world's most important center for quantum reality research at the Institute of Theoretical Physics in Austin, Texas, would agree with Siegfried. In his early years,

Wheeler believed that the universe was composed of particles and fields. Now, however, Wheeler believes that the most basic building block of the universe is information: "I have been led to think of analogies between the way a computer works and the way the universe works. The computer is built on yes-no logic. So, perhaps, is the universe" (Siegfried, 5). Some have taken this idea of a nonmaterial universe one step further. Physicist David Bohm (1917-1992) has concluded that the universe is organized at all levels of complexity according to *meaning* in what he calls the "implicate order." "Wholeness" is more than a theoretical construct according to Bohm. We are part of the undivided whole, and since there is no division between mind and matter, there can be no boundary between science and subjective human experience (Bohm, 5).

The above experiments give us a solid foundation about the true nature of reality. We can see from these experiments that the world is a much stranger place than most of us imagined. Many questions remain to be answered, and I, for one, cannot think of anyone who would be more qualified to resolve these questions than Albert, Niels, and John. Let's meet them now.

Albert, Niels, and John

THE DATE IS April 18, 2030, one hundred twenty five years after the publication of the theory of special relativity. Albert Einstein has been invited back to Princeton University where he lectured in 1921. In 1933, he had moved to Princeton permanently and worked at the newly founded Institute for Advanced Study and Palmer Physics Laboratory, now called First Campus Center. Albert has been given the office he occupied from 1933 to 1939 at 109 Fine Hall, now called Jones Hall. Niels Bohr and John Stewart Bell are scheduled to join Albert shortly and will take residence next to Albert's office.

Albert, Niels, and John, were "resurrected" in 2024 by a method or technology whose details are beyond our grasp and are not important for our story. All we know for certain is that these three men have been "brought back from the grave," and they are healthy and vigorous. For the last six years Albert, Niels, and John have been trying to catch up on the latest scientific, technological, political, social, and cultural issues. Albert has been living and studying in Zurich, Switzerland; Niels in Copenhagen, Denmark; and John has been in Berkeley, California. All three are looking forward to getting together at Princeton to engage in discussions on the meaning of quantum theory. In particular, they have been invited back to participate in analyzing the data from an experiment called the *Bell Project* named after John Stewart Bell. The data from the project will be sent to Princeton University from NASA headquarters in Pasadena, California.

Over the past two years NASA has launched three separate probes into deep space. All three probes have been deemed successful and are comfortably in their respective positions, prepared to carry out the designated experiments to confirm John Bell's theorem of inequality. While laboratory experiments were performed in the 1980s and 1990s that verified Bell's theorem over a distance of several miles, this will be the first test to verify his theorem over distances of millions of miles. Though most are convinced of the validity of Bell's theorem of inequality, a few skeptics remain, including Einstein himself.

As for its three esteemed and honored guests, Princeton University has gone to great lengths to make them feel welcome. Their offices have been refurbished with special care given to Albert's office to give it the look and feel it had when he first moved to Princeton in the 1930s, save all the new comforts and conveniences. At that time, he was offered a larger office, but he preferred the smaller, cozier confines and delegated the more spacious office to his friend Niels. Let's pick up our story and eavesdrop on the conversations of Albert and Niels upon their arrival at Princeton.

Albert is sitting in his office chair at his desk pouring over some of the technical issues and experimental procedures for the upcoming Bell project when he hears a knock on the door.

ALBERT: Come in.

NIELS: Albert, how have you been? They told me you arrived earlier this week. It's great to see you again!

ALBERT: You're looking great, Niels. How was your trip?

NIELS: Good. I just got in a few hours ago. I haven't settled in yet, but I just had to come over to see you. I can't wait to get started with this project. This is going to be fun.

ALBERT: It's just like old times isn't it, Niels?

NIELS: Ya, the only differences are that we have all these new gadgets and gizmos to play with. I see that you've got your computer on. Have you figured that thing out yet?

ALBERT: I'm slowly getting the hang of it. I still can't believe how fast it is. All of that information on a tiny silicon chip. It's just unbelievable. We would have never imagined this in our day, would we?

NIELS: No. And we owe it all to the quantum. It doesn't seem like all that long ago that the two of us were standing right here in your office. So, much has changed, and yet so many things are the same. I thought people would be maneuvering around in their little personal flying machines by now, but the traffic on the roads is terrible. Have you been out there? It's crazy.

ALBERT: Well, as you know, I never learned to drive in my first life, but I thought that it was time to learn, so while I was in Zurich, I practiced and passed the driver's test. I took another test when I got here and now I have a New Jersey license. Princeton officials saw to it that I have a car, but I only use it when I leave town. I usually walk or take the bus.

What really surprised me is how little cars have changed since the days of Henry Ford. They're still basically a carriage on four wheels with a gas guzzling engine that spews toxic emissions. Why did people ever think that millions upon millions of these machines emitting carbon dioxide would have no effect on the environment? I read the other day that all of the ice completely disappears in the Arctic in the summers now. The polar bear is gone, the seals, all the exotic species that lived in the sea below the ice. It is so sad.

NIELS: I guess we're all to blame really. We've used all the resources on the planet as if they would never run out. Let's face it; the real problem is that there are just too many of us.

ALBERT: Ya, something has to give.

NIELS: When will John be here?

ALBERT: Not for several weeks. He's working on some kind of project out in Berkeley that he can't tear himself away from right now.

NIELS: Ah, too bad. I'm looking forward to seeing him. What have you been up to since you arrived?

ALBERT: I've been looking over some of the two-hole experiments that have been conducted over the past fifty years or so. I remember that Thomas Young did the first experiments in this regard in the 1800's when he passed light through a tiny hole and noticed wave interference. He concluded that light was a wave, but then, as you might remember, I came up with the idea that it was a particle just as Newton had imagined. Turns out Young and I were both right.

NIELS: Yes, and for that you got the Nobel Prize in physics. What did you do with all the money they gave you?

Albert doesn't realize that Niels is teasing him and proceeds to offer an explanation.

ALBERT: Mileva, my first wife got it. At least she got all the interest from it. It was part of the divorce agreement.

NIELS: Who came out ahead on that deal?

ALBERT: It was mutually advantageous. I did it mostly to support my two sons. You know Mileva and I reconciled in our later years. She did a wonderful job of raising Hans and Eduard.

Albert gets back to the business at hand.

ALBERT: Niels, they've done these experiments with all kinds of subatomic particles, electrons, photons,—you name it. Take a look at this, Niels. Here are some illustrations of the two-hole experiment with photons of light. What do you think?

Niels takes the paper from Albert and looks at several experimental set-ups concerning the two-hole and photon experiments. (Refer to chapter3 on the two-hole experiments) He is already familiar with these experiments, so he just takes a cursory look.

NIELS: It's all very odd isn't it? But this is as I had predicted. No subatomic event is a real event until it is observed.

ALBERT: I'm still not entirely convinced that nature can be so fickle. I'm wondering if there isn't some other explanation possible here, maybe something involving my relativity theory and higher dimensions.

NIELS: Well, I'm not surprised you have some other explanation. Let's hear it.

Niels smugly fills his pipe with some tobacco and draws a match not wanting to appear too interested in Albert's explanation.

ALBERT: Now, you claim that the electron exists in a superposition of states. You also claim that the electron has no reality until it is observed, is that correct?

NIELS: Sure.

ALBERT: And would you also agree that what you refer to as the "observer effect" really isn't an actual physical effect. After all, one can't "will" the particle to go through one hole or the other. There is always a 50/50 chance of going through one hole or the other. In other words, the particle goes wherever it pleases with a 50/50 probability. It's nature's choice.

NIELS: That's pretty much correct.

ALBERT: Ok, and you seem to be impressed with the idea that the electron in the two-hole experiment seems to "make a decision" as to which hole it will go through if there is going to be an observation in its future. If there is no observation about to take place on the opposite side of the hole, it will go through both holes simultaneously, but if there is going to be an observation, the particle will choose one hole or the other. Do I understand that correctly?

NIELS: Yep, but remember, it doesn't actually go through both holes; only its statistical wave goes through both holes. Your friend Max Born was the first to demonstrate this aspect of the quantum in the 1920s. Light itself is a statistical wave. There is no reality to the photon or the electron from the time it leaves the source until the time it is observed.

ALBERT: I'm not sure I buy the *statistical* wave thing. Are you saying that when I look up at the stars at night, the light does not travel in an electromagnetic wave between the star and my eye? Are you saying that it is only a statistical wave and that the photon only comes into existence when I perceive it?

NIELS: Exactly.

ALBERT: Well, even if what you say is true, it's a moot point and here's why: Relativity states that, for the electron or the photon traveling at the speed of light, time and distance shrink to zero. No time passes for the photon in flight. The photon has no history, it has no "before" or "after." The dimensionality of space and time in the experimental set-up is only relevant in our world of dimensionality.

Let's take the two situations, one with an observation and the other without. When no measurement occurs, the electron goes through both holes and continues on to the screen. Remember that the two holes are extremely close, and it doesn't seem out of the question that the electron wave could physically pass through both holes at the same time. Now suppose we install the detector at one of the two holes. The electron wave has a 50/50 chance of being absorbed by the detector and a 50/50 chance of not being absorbed and traveling on to the screen. But the reason for its being absorbed or not being absorbed doesn't have anything to do with your supposition. It all depends upon where the particle is located in the wave. If an observation is set up at hole A and the electron within the wave packet happens to be closest to hole A as it arrives, then the electron wave will collapse as its energy dissipates and the detector will record a hit. But if the electron happens to be on the opposite side of the wave packet as it gets to the holes, it will pass undetected through hole B and continue on to the detection screen, its wave undisturbed. It can all be explained classically.

NIELS: Al, I can see where your explanation would be possible in the two-hole experiment with the electron because of the fact that the two holes are in such close proximity, but it won't work in the second experiment with the photon. Path A and B can be separated by any distance—a meter, or a kilometer, it doesn't matter. The results will be the same. There is no way that a classical wave can be spread out that far. If the unobserved particle is a probability wave, as I believe, then yes, the statistical wave can theoretically spread across the universe, but a classical wave cannot do this. The photon experiment clearly rules out the idea of a spreading classical wave. And as far as your Relativity explanation that no time passes for the photon, I must agree with that. No "before" or "after" exists for the photon, but there certainly is for the experimenters, who still measure the photon traveling at 186,000 miles per second. For them, there is a "before" and "after." If you are unwilling to give up your idea of a classical wave, then you must face the consequences of your own relativity theory, which means you must explain how it is possible that the photon is being affected retroactively. When a detector is set up in one of the photon paths, well after the photon should have already proceeded down one of the two paths, the results dictate a backwards-in-time scenario if you maintain that we are dealing with a classical wave. This is in violation of Relativity, because Relativity asserts that within a time/space reference an effect cannot precede its cause. The experiments prove this out, Al. We have to either give up Relativity, which neither of us is prepared to do, or you must admit that the electron and the photon are only statistical waves—waves that have no physical reality.

ALBERT: It doesn't necessarily have to be one or the other, Niels. One possibility is that we are dealing with more than three spatial dimensions. When I was working on my unified field theory in the 1920's I ran across the work of mathematician Theodor Kaluza and physicist Oskar Klein, who were playing around with the mathematics of four spatial dimensions in order to describe my space/time coordinates called metric tensors. Both wondered if these mathematical devices might have some physical reality. At the time, I was critical of their ideas because no evidence of extra dimensions existed, but, in retrospect, I'm not so sure. Remember that both Planck and I thought that his mathematical formulations describing the radiation emissions from a black-body were not rooted in reality. Then, as you remember, I used the same mathematics to describe the photo-electric effect. Like Planck, I was not sure that this was not anything more than a mathematical trick, but it turned out that the math described physical reality perfectly.

If math is as objective as it seems, it is quite likely that Kaluza and Klein were on to something. Their mathematical models could well describe physical reality. In fact, the space/time curvature around a three dimensional body requires five dimensions—four spatial dimensions and one time dimension. It is impossible to represent the curvature of space around a three dimension object. Go ahead, Niels, try to imagine space curved around the earth. Try to draw that situation. It can't be done. You can't imagine it or draw it in three dimensions; therefore, at least five spatial dimensions must exist.

So, it's quite possible that in the case of these particle experiments the particle resides in an unseen dimension until a measurement is made. This to me seems like the more probable scenario. Why drag in some kind of statistical wave hocus-pocus when we can deal with physical reality: a reality that we can measure, not directly perhaps, but through its physical effects on space/time.

NIELS: Al, attempting to pull a fifth dimension out of your hat to save an argument seems a bit desperate. No extra dimensions have ever been detected and as a positivist, I'm surprised you would attempt to make such a radical assumption. If it can't be measured, there is no reason to speculate blindly. All we can say is how nature behaves. We can't say what nature "is."

ALBERT: You have no room to talk, Niels. Your idea of the statistical wave and the collapse of the wave function are immeasurable attributes and they have no physical reality. You maintain that the collapse isn't really "caused" by an observation. In fact, you really don't offer any tangible reason for the collapse, and yet, you accuse me of making wild speculation about five dimensions. At least my speculations involve physical reality. The only thing that is going to save your idea of the collapse of the wave function by measurement or observation is if we broaden the definition of cause and effect to include an affect by the observing person or apparatus. If the definition of causality is broadened to include a physical collapse of the wave into a single event, then we could get back to a classical explanation, one in which we aren't faced with this damned statistical wave. But if you aren't ready to admit some kind a causal explanation and you stick with your observe- created reality scenario, then we're faced with the uncomfortable situation of a non-physical consciousness collapsing a non-physical wave. I don't think you are really going to reduce yourself to such nonsense.

NIELS: No, I'm not prepared to say that at this point. We just don't know what causes the collapse of the wave function, if in fact that is what happens. We don't know if it's the measuring apparatus or consciousness itself. We don't know why an observation manifests a change from a wave to a particle. That, in a nutshell is the quantum enigma.

At that moment there is a knock at the door. A promising young physics student by the name of Rashid has been chosen to look after Princeton's esteemed quests and cater to their every wish. Many applied for the position, but the brilliant young graduate student was chosen due to his high academic achievements. He is delighted that he has been given the opportunity to meet these famous men and is eager to make a good impression.

ALBERT: Come in, please.

RASHID: Excuse me; I hope that I'm not interrupting anything. My name is Rashid, and if there is anything at all you need, please let me know. I'm in room 115 just down the hall. I'm working on some of the logistics of the Bell Project, and as a result, I'm practically living in my office, but I always have time for you Mr. Einstein and Mr. Bohr, so here is my phone number. You may call any time day or night, and I'll be here to answer any questions or run errands. Don't hesitate to call or just knock on my door.

Rashid hands the number to Albert and shakes his hand. Rashid bows slightly as if he is meeting royalty. Niels comes over and shakes the lad's hand.

NIELS: Yes, we'll certainly do that if we need anything, but please, just call us by our first names, Albert and Niels.

RASHID: Well, I should go and leave you to your conversation. It was so good to meet you both. I'll see you again soon I hope.

ALBERT: Assuredly.

Rashid leaves, and the two get back to their conversation about the two-hole experiment.

ALBERT: It is just all too strange. It's hard to know what to make of it. The strangest part of all is the apparent backward-in-time affect. How does one explain the fact that the interference pattern disappears when the observation takes place after the particle already passed the point of no return—after it has supposedly already gone down one path or the other. It's true that relativity theory says that a cause cannot precede its effect, but relativity theory also suggests that the future is every bit as real as the past. If we could somehow bring in relativity to show that there is some kind of signal from the future, then we could get back to some kind of sensible determinism. That would get rid of the anthropocentric notion that human consciousness is responsible for the wave function collapse.

NIELS: Whoa! I can't believe you are talking about "sensible determinism." Weren't you the one who once said that common sense is the set of all prejudices we attain by the age of eighteen? And as far as determinism goes, certainly you don't think that such things as atomic decay or the jump of the electron from one orbit to the next is determinate, do you?

ALBERT: Yes, I do. The systems as a whole are determinate. For example, if you have 20 lbs. of a radioactive element with a half-life of 10 years, you know for certain that only 10 pounds will be left after 10 years. This is a basic deterministic law of nature.

NIELS: Sure, I agree that the system as a whole is determinate, but it is not determinate for the individual atoms. Each atom decays by pure chance. There is no timer telling the individual atom when to decay. And this is the very mysterious part of the whole thing. If you have a sample of an element, regardless of the size, precisely one half of it will be left after its established half-life period. It makes no difference how you treat the material either. You can heat it, cool it, pulverize it; it doesn't matter what you do, the result is the same. You might have a sample with trillions of atoms or a sample with only 100 atoms, after its known half-life, one half of the element will decay in each sample in a purely random way. How does each atom of the sample know when to decay so that it always maintains the exacting half-life?

ALBERT: Remember that Aristotle had an interesting idea called teleology. It is about "becoming." An oak tree coaxes the acorn into becoming a tree. As you know, relativity theory states that the future and the past actually exist in reality. The future and the past are every bit as real as the present. Isn't it possible that nuclear decay can be understood in terms of teleology?

NIELS: What do you mean exactly?

ALBERT: Well, as you stated, Niels, with nuclear decay, we know that half of the decaying element will remain after a given amount of time no matter how much of the element one starts with or how it is treated. This only makes sense if the situation is temporally reversed. The only way that one-half of the atoms in the substance will decay to such a precise amount every single time is if the end goal is known in advance. There must be some kind of signal from the future.

Physicist Julian Barbour and Gevin Giorbran each produced works that I believe are relevant to the topic we're talking about. Newton believed that space was objective. All motion, he assumed, moved relative to absolute space, but as you know, I demonstrated that space is a relative concept. One must consider both space and time together, and I synthesized the two into space/time which is an objective coordinate. But Barbour felt that I had not gone far enough, and he went one step further by eliminating the concept of time altogether. In order to achieve this feat, he also had to rid us of space and motion, which is no easy task and very counterintuitive to boot.

NIELS: You've made a living being counterintuitive.

Albert takes a long draw on his cigar, seemingly relishing Niels' comment before he continues.

ALBERT: The way Barbour handled this idea is to say that the universe is nothing more than the objects that comprise it. Space, he said, is nothing at all; it's just the distance that separates the objects. Space merely defines the separation of objects like the spaces between words in a sentence. So in essence, what Barbour is saying, is that the universe **is** the objects, and there is nothing outside of the universe in which to judge movement. There is no "God's-eye view"; The objects are all that exist, and so what we think of as motion is nothing more than different configurations of these objects. There is no motion, just different snap shots of the universe. In this way he got rid of the concept of motion.

NIELS: Interesting. But what's your point regarding nuclear decay?

ALBERT: Hold on, I'm getting there. I really think Barbour's idea is inspirational, and I very much like the utility of it, but there is a problem. Many of nature's laws have been challenged over the years, but one law that seems to be in solid bedrock is Boltzmann's second law of thermodynamics. It appears unquestionable that usable energy diminishes as systems move toward entropy. Boltzmann was wise enough to see that his principle had implications for mathematical laws of probability. Matter and energy move toward entropy because this is matter and energy's most probable state. Boltzmann maintained that entropy and probability are intertwined, and this I believe was his greatest contribution.

The second law of thermodynamics is what gives us the arrow of time. And this, I think, is the shortcoming of Barbour's idea. I don't believe that different configurations of the universe account for this perceived arrow of time. He doesn't account for the thermodynamic process of entropy, the drive toward the most probable state. Expansion of the universe is not a discrete proposition, but rather a smooth and continuous affair with the tug of gravity creating energy from the expansion of space. Space, I believe, *is* something real; it has texture, and I'm not so certain it can be whisked away as Barbour would have it.

So along comes Gevin Giorbran. His idea is a bit more comparable to mine. His universe includes an Alpha and an Omega, a beginning and an end. While I believe that both the past and the future exist in reality, he believed that the Alpha and the Omega exist in potentia, both being constantly created from observations in the present, analogous to John Wheeler's participatory universe idea, in which an observation in the present influences the path of a particle in the past, and similar to what you were talking about with the two-hole experiment. In some sense Giorbran's idea allows for a "God's-eye view", not from outside the universe but from within. The past and the future actually exist as in my block-time universe idea.

To conceive of the block-time universe one must first imagine a two-dimensional plane moving through space and carving out three dimensions as it moves. Now, extrapolating, imagine three dimensions moving through space and tracing out the fourth dimension of time.

But I think Giorbran's most unique idea is that time is fundamental. Giorbran's idea of time is a bit like Barbour's idea of space. Just as Barbour maintained that space is nothing more than the separation of objects, Giorbran says that time is a manifestation of the utilization of energy in the transformation of matter into the symmetry of space. Time, like Barbour's idea of space, has no intrinsic value in and of itself but is an epiphenomenon of the relationship between energy and matter. So, even though Giorbran doesn't believe time in and of itself is anything real, it is nevertheless fundamental. It is a by-product, an entity if you will, of the process involved in the universe's attempt to find balance. He would contend that the universe is not winding down, as the typical understanding of the second law is understood, but that the universe is transforming from one type of order he calls "grouping order" to another type he calls "symmetry order" (Giorbran 61).

Although Giorbran doesn't explicitly state his concepts in this manner, I believe that his is a philosophy of the primacy of mathematical law. If

I understand his ideas correctly, he denies that there are any "forces" in nature. He maintains that space is denser than matter, and he uses this idea to convey his belief that the apparent gravitational force around massive objects is the result of dense space trying to fill the relative vacuum of matter. We could envision an air bubble representing matter as it moves through a more dense fluid representing space. The fluid presses in on the air bubble from all directions, creating a near perfect sphere of air. This idea would serve as an alternative to my earlier idea requiring five dimensions. This would solve the problem of how space curves around matter in only three dimensions. I actually prefer his idea because of its simplicity.

Likewise, the supposed electromagnetic "force" is simply the result of particles trying to attain symmetry or balance. There is no attractive electric force between the electron and the proton; rather, it is nature's attempt to find mathematical balance, the most probable state of symmetry, the zero state. The electron's negative charge (-1) is balanced by the proton's positive charge (+1) to create a balance of zero. The early universe began in perfect symmetry, the alpha state, but soon after the big bang, asymmetry developed as matter began to clump together to form what Giorbran called grouping order. Grouping order is a probabilistically-unlikely state, and as the universe evolves, it moves toward a more probable omega state of perfect symmetry. We can understand these forces in terms of objective mathematical law. The probability he is talking about, I believe, is not the random chance of quantum theory, but a deterministic probability driven by the objective mathematical laws of nature.

NIELS: So what does this have to do with nuclear decay?

ALBERT: Well, you were wondering how it is that an individual atom knows when to decay in order to reach the half-life state in which only one-half of the original sample remains. The problem here is that you are thinking of time as something objective. You know as well as I that time is nothing more than motion. For our purposes here, time is nothing more than the process. It is the process of nuclear decay in and of itself that creates time. One might say that the second law of thermodynamics does not necessitate time; time necessitates the second law. The same holds true of nuclear decay. The individual atoms need no external clock to tell them when to decay. So if we use a combination of Giorbran and Barbour's idea, we can make sense of this nuclear decay thing:

Suppose that you have one hundred pennies representing a sample of a radioactive substance. All the pennies lying on the table show "heads," which is an extremely unlikely state. It would be much more likely if about half of the pennies were heads and the other half tails. The state of all heads represents the improbable state of the early universe. Now suppose that all of the coins that are heads are able to spontaneously flip over to tails in order to obtain a more probable state. Each coin has an equal probability of flipping to a state of tails during a certain configuration of the universe. Initially, from a "God's-eye" perspective we would expect a large number of flipping incidences when all of the coins are in a state of "heads", just as when we have a large sample of a radioactive element, a Geiger counter will report many decaying particles. But remember, there is no external clock. The events themselves are the clock. Each spontaneous flip of a coin creates a new configuration of the universe. As the universe moves toward the omega state we get closer to achieving balance. Eventually, the sample will reach its half-life of about 50 heads and 50 tails. But as you see, there is no objective clock in which the individual penny must abide to reach the end goal. It is the process itself that creates the time clock of the half-life, not the other way around. And since the universe is a quantum universe, we can say that the process of radioactive decay, which is a quantum event, is the standard clock of the universe. There is no reason for you to wonder why some atoms, or coins in our example, decay right away and some wait for years to achieve your notion of half-life. The atoms are just trying to find balance as they move to their most probable state. Each new configuration is a new roll of the dice; each new configuration of the universe resets the universal clock.

Of course my analogy with the pennies isn't a perfect one. In the case of the coins, equilibrium was achieved when they reached a state of 50 heads and 50 tails, but equilibrium would not occur in the case of a radioactive substance. It continues to decay until there are almost no radioactive atoms left in the sample.

But I want to be clear here. Chance does not explain why things happen. Chance and the laws of probability only explain what happens—not the reason why it happens. The reason is a deterministic one. It is Boltzmann's second law of thermodynamics. And in his book *Cognition and Chance,* Nickerson reiterates this idea:

> *In my view, chance is a descriptive concept and never an explanation. We may say that under specified conditions, certain events are best described as chance events—that the various possibilities have equal probability of occurring—and this observation may be very useful, but simply calling it chance does not explain why the behavior is such as it is. The atoms of a radioactive substance are constantly decaying, the rate of decay differing over a very large range from substance to substance. We say that precisely which of the atoms will decay during any particular period—or whether any specific atom will do so—is a matter of chance. But by calling it a matter of chance adds nothing to the observations that every remaining atom has the same probability of decaying immediately, and that this does not change over time. And the invocation of chance does not explain why the decay process is as it is.*
>
> *Chance is one of those terms that lend themselves to the fallacy of reification—the uncritical assumption that a name must name some thing. In invoking the name, we convince ourselves that we have explained some phenomenon of interest. But in what sense have we done so?...The inverse square law of gravitational attraction appears to be a universal law, something that characterizes the behavior of matter everywhere and everywhen and the observation that this is the case is immensely useful, but stating the law does not constitute an explanation of the why of it. Just so with the concept of chance.* (Nickerson, 52)

So you see, Niels, each time an atom decays within a sample, it's a new configuration of the sample and a new configuration of the universe as a whole. The universe has moved a bit closer in time to symmetry, which in turn will nudge the next atom to decay in the succeeding configuration. One cannot talk about an atom from a previous configuration waiting for years to decay, because each configuration restarts the process anew. There is no history since according to Barbour time and motion do not exist—only new configurations of the universe.

So from the teleological standpoint, it isn't so much that there is a signal from the future as there is a mathematical goal to reach a state of equilibrium: the second law of thermodynamics. A modified second law to be sure, but it's the second law nonetheless. This is the most probable state; the Omega state of perfect symmetry that drives the process. A past and a future exist, but it changes form with every new configuration of the universe. This eliminates the anthropomorphic necessity for an independent observer to create the past and the future as Wheeler would have it. As you see, I am using both Barbour's and Giorbran's ideas of the nature of the universe to explain the oddities of nuclear decay that I have found so troublesome.

In Barbour's universe, time and motion do not exist, so one cannot speak of a history. In Giorbran's universe, time does exist, but it is driven by the second law of thermodynamics: the drive toward balance. Like kernels of popcorn waiting to be popped in a skillet on the top of the stove, each decaying particle moves toward balance as a result of its inevitable drive toward symmetry. Just as all the kernels in a kettle don't pop all at once, so each atom of the element waits until the conditions are right for decay. As the pull of symmetry tugs on the atoms, they reach the point of decay, and that in turn nudges the next element into decay. The driving force, the engine if you will, is the deterministic second law of thermodynamics.

NIELS: Phew, Al, that is really a stretch, but I commend you. Somehow you have succeeded in making the process of nuclear decay determinate. I see two problems however. First, with your scenario wouldn't we see a burst of radioactive decay if the configurations are discreet snapshots, rather than random individual decaying particles evidenced by the click of a Geiger counter?

ALBERT: Well, with Barbour's idea that is somewhat of a problem, and that is why I like Giorbran's idea a bit better in that the smooth continuous arrow of time is preserved as we move to a state of symmetry. What we need is a synthesis of Barbour's and Giorbran's ideas. I may spend more time trying to synthesize these two ideas.

NIELS: That doesn't surprise me.

ALBERT: Niels, if I had a coin and flipped it twenty times would you be surprised if it came up heads every time?

NIELS: Sure. I would think that you were cheating somehow.

ALBERT: No, you would have just witnessed the strangeness of classical chance. You might well be surprised if I had gotten twenty heads in a row but twenty heads in a row has the same probability of coming up as any other series of twenty outcomes. Each flip has a 50/50 probability of a heads or a tails. Having flipped nineteen heads in a row doesn't make it any more likely that the twentieth flip will result in a tail.

Suppose that whenever you roll a "six" with a die, an atom will decay. And suppose that you have rolled the die twenty times and you still haven't rolled a "six." The atom still has a one-chance-in-six of decay on the next roll. This is what makes atoms of a radioactive element appear ageless in the quantum sense. In each roll, or each configuration of the universe, each atom of a sample still has the same likelihood of decay as it did from the very beginning. It always maintains a one-chance-in-six of decay no matter how many rolls of the die have occurred previously. No need to deal with spooky quantum chance; as surprising as it may be, this is nothing more than pure classical chance.

NIELS: I like your idea, Albert, but in your Giorbran/Barbour universe the rate of events wouldn't remain constant over time and, therefore, the number of configurations would change over time. For example, when the universe is small and gravity is intense, I would think the Giorbran's idea of grouping order would dominate and the drive toward entropy would be very slow, but as time goes on and the universe expands exponentially, the pull from the future increases and many more events happen than previously. If events are what drive time as you maintain, wouldn't the universe's expansion increase the number of configurations and rolls of the cosmic die? Wouldn't this tend to speed up the rate of nuclear decay? We've seen that the decay rate is very stable over time. How would you explain that?

ALBERT: Ok, well, imagine that we have a movie camera that represents the universe at large. Let's call it the Barbour/Giorbran camera. Suppose that you would like to observe the extremely fast fluttering of the wings of a hummingbird. To achieve this you could take a movie of the hummingbird with the camera running at high speed. That has the effect of slowing down motion for observers who watch the movie when it is played back at normal speed.

Now, suppose that while the humming bird is hovering, the Barbour/Giorbran camera is filming the humming bird, an observer, and a clock hanging on the wall in the background. From the camera's perspective and individuals watching the film being played back at normal speed, the motion of the hummingbird's wings and the hands of the clock appear to have slowed. For the subjects being filmed within the Barbour/Giorbran universe however, nothing has changed. The hummingbird's wings are still flapping at the same rate as before because the clock has slowed proportionally. If you were to count the number of flaps per second according to the slowed clock in the background you would find that there is still the same number of flaps per second as before.

This phenomenon is what gives radioactive decay such a stable clock. It is true as you say that as space and time expand in a universe that is accelerating toward omega, events seem to speed up from the "God's-eye" perspective. We could call this the objective clock. But one must remember that, if space/time expands, then time for individual events must slow to compensate for this expansion of space/time. And since events are what drive time, as we determined earlier, the number of events per unit of time remain the same. Just as in the Barbour/Giorbran movie in which the hummingbird's wings seemed to have slowed from the perspective of our clocks, the same number of flaps occurred per second according to the clock in the movie. We judge the rate of atomic decay from our vantage point within the universe in which we live. From a "God's-eye" perspective the expanding universe does seem to speed up the rate of events, but from the perspective within, from the clocks within the universe, nothing has changed. When we determine the half-life of an element, it is determined according to our personal clocks in the here and the now.

Niels, I have a puzzle for you to solve. If you woke up one morning and found that everything had doubled in size would you notice?

NIELS: Everything? Would distances double as well?

ALBERT: Yes, of course.

Niels ponders the situation for a few minutes and then begins to think aloud as he is accustomed.

NIELS: Hmmm. Well, let's see. Since everything was twice its normal size, it would all be proportional. I would be twice as large, but so would my house and all of its surroundings. If I pulled out a ruler and measured things, they would appear the same size because the ruler had doubled in size as well, so I wouldn't be able to determine that anything had changed by using measuring instruments. If I looked up in the sky, the sun and moon would look the same size as they always do because, although they are twice as larger, they are also twice as far away so their apparent size would look the same to me.

ALBERT: You haven't lost your sharp mind, Niels.

NIELS: Hmmm. I've got an idea. In 1969 the Apollo astronauts left a mirror on the moon so that a precise measurement of its distance from earth could be measured by bouncing a laser beam off the mirror and noting the time it took for the light to make the round trip journey. Since the speed of light is always the same according to your special relativity theory a precise measurement of the distance can be made. So, I would fire a laser beam at the mirror on the surface of the moon and record the time it took to get there and back. It would take twice as long for the round trip, so I might surmise that since the apparent size of the moon looks the same as it always has, its distance and size must have doubled.

ALBERT: Good try, Niels. But you get an "F" in Relativity. You are right that the speed of light is a constant, but time and distance are relative. Remember that your meter stick has doubled in size as well, and the speed of light is measured in terms of meters per second, or miles per second. If the distance doubles, time must slow to compensate. Light always travels at 186 thousand miles per second in a vacuum. To compensate for the doubling of the distance, the duration of a second must double as well so that one still gets the same result when calculating the speed of light. It would still take the same amount of time for light to travel the distance from the earth to the moon and back again when the distance doubles. Time has slowed proportionally to the doubling of the distance.

One might be able to detect that everything had doubled in size in other ways, however. For instance, you could stand on a scale and find that you weighed approximately four times as much because both the earth and you would have doubled in size and as a result the gravitational force would almost quadruple.

NIELS: Yes, I see what you are getting at. First, making a fool of me, and second demonstrating that events, and therefore by definition, time would slow in relationship to an expanding universe. This would indeed allow for the stable clock of radioactive decay. Very impressive. You really thought this one out. But, my head is beginning to hurt and I'm getting very tired. What do you say we pick this up another day?

ALBERT: Good idea. I wouldn't mind turning in myself. I sort of got off the subject didn't I?

NIELS: No problem, I'm glad we got off into that discussion. Let's get together toward the end of the week for breakfast? Do you know of any good places?

ALBERT: There's a little café in the campus towers, we could go there.

NIELS: Where's that?

ALBERT: Up on Prospect Avenue, less than a block away. How about around 8:30 Friday?

NIELS: Sure, that would be great. See you then.

Albert and Niels hear a knock on the door. It is Rashid.

RASHID: Sorry to bother you again. My girlfriend and I ordered a couple of pizzas and we were wondering if you two would like to come over and have some pizza with us?

NIELS: I was about to call it a day, but how can I turn down pizza and good company. What do you say Albert?

ALBERT: Ya, sure. We'll be over in a few minutes.

RASHID: Great! Risa and I will be honored. The pizza should be here in a few minutes, so we'll see you soon, ok?

NIELS: We'll be there shortly.

After a bit, Albert and Niels arrive at Rashid's office where they meet Rashid's girlfriend Risa, an attractive, tall, slender, blond woman, who towers over Rashid by a good three inches. Risa is a journalism student and has been assigned to cover the developing story of Albert, Niels, and John for the University student newspaper. She is also keeping a journal as she is in the midst of writing a historic fiction concerning the lives of Einstein, Bohr, and Bell. Progress on the book is going well, and she is eager to pick up some more anecdotal stories before the book goes to publication. .

Albert and Niels find some make-shift seating and enjoy some pizza and small talk with their new friends. After the feast Rashid and Risa ask their two guests many questions about their former lives and the people they knew.

RASHID: Albert, What was it like growing up in Germany? What were your parents like?

ALBERT: Well, as you know I was born into a Jewish family, but my parents were irreligious, so we didn't practice the faith. We weren't rich, but we never lacked for the basic things in life. I guess we would have been considered middle class. I was born in 1879 and my sister, Maja, followed a couple of years later. We were very close throughout our lives, but my parents liked to remind us how we fought as kids. It's surprising that my sister survived my tantrums. It seems she was constantly ducking projectiles that I threw at her. She had very good reflexes.

RASHID: Is it true that you didn't speak until you were four years old?

ALBERT: Yes, that's right. If I had been born in this century they would have labeled me autistic. Perhaps it was because of my inability to speak that I thought in pictures rather than words, and this ability served me well in later life.

RASHID: How did this affect you in early life?

ALBERT: I was considered odd, and slow by my family and friends. They began to worry that I had a mental disability. I kept to myself mostly and found it very difficult to bond and make friends. I withdrew to my own little world. It was much more interesting to play with objects than play with other kids. I loved mechanical devices and I loved to take apart watches to see how they worked. I often spent hours totally engrossed in building towers with playing cards.

RISA: Did you break out of your inner world?

ALBERT: Yes, music opened my world. My fondest recollection of those days was when mom would play the piano, and we would all sing. She saw how music brought me out of my shell, so she arranged violin lesson for me. We often played duets, her on the piano and me on the violin. Music made me more passionate, not just about music, but about life. I especially loved Mozart, because it seemed the music came from the great beyond. I felt that Mozart didn't create the music so much as he channeled it. My violin was not only my musical instrument but a tool that opened up the universe to me. Whenever I became stuck on a problem, I would pick up the violin and play, and then suddenly, as if from nowhere, I would have a revelation and solve the problem.

RISA: Was this the beginning of Albert the determinist?

ALBERT: Yes, I would have to say it was. Often the music and the knowledge seemed to come out of the void. It is like the riddle of the sculpture. Does he chisel away on the rock to create a face, or does he remove the surrounding rock to discover the face inside? This is one of the great questions about mathematics as well. Is it discovered as physicists like Roger Penrose believes, or is it invented. I tend to think the former.

RISA: How about you Niels? What was your early life like growing up in Copenhagen?

NIELS: I was born about six years after Albert. In many respects our families were similar. My parents, like Albert's, were nurturing, secular, social progressives, and we both had parents who understood the value of education. It was a bit easier for me to attain an education, however, because I came from a long line of academics. My father was professor of physiology at the University of Copenhagen.

I had two siblings. My sister Jenny was a year older, and I had a brother, Harold, who was about two years younger than I. Harold and I were very close and I always considered him my best friend. He did everything better than me. He was brighter, he was a better footballer, and he graduated from college almost a full two years before I did. He played football on the Olympic team and brought home a silver medal for our country. But there was never any competitiveness or jealousy between us. We always rooted for each other.

Albert and I had another thing in common. I was considered odd as well.

RISA: How so?

NIELS: People in my home town love to tell visitors the story of when Harold and I were playing football, which they call soccer here in the United States. In one particular game against an arch-rival German team, I was playing goalie and my team was on the attack at the opposite end of the field. With no immediate threat from the German team, my mind started to wander and I began to think about a math problem I had been working on. I started to scrawl some formulas on the goal post and about that time the German team had stolen the ball and everyone was racing down the field toward our goal. One of my teammates noticed I was paying no attention and yelled at me just in time for me to regain consciousness and block the ball. I was totally engrossed in that mathematical problem (Spangenburg, 1).

RISA: Are there other examples you can tell us about?

Risa is furiously taking notes, not wanting to miss any information for her story.

NIELS: One day when I was very young my father and I were out for a stroll and, in a teaching moment, he tried to draw my attention to the symmetrical geometry of a tree. He thought it miraculous how the trunk of the tree extending into the branches, and the branches into twigs, and the twigs into leaves. And I said, "Of course, because if it weren't that way it wouldn't be a tree would it?" (5) Like Albert, I just tended to see the world a bit different from others, and I believe it is because of this that we were able to break the mold as it were.

RASHID: That is intriguing. Albert told us that he didn't speak until he was four years old. Did you have anything similar in your childhood?

NIELS: Yes, yes. My problem with language was a bit different however. I could speak alright, but I always had trouble organizing my thoughts and this problem plagued me my entire life. When I was in college I had to rely on my friends to help write papers. My fiancé at the time, Margrethe Norlund, helped me with my college papers and my thesis. It was very difficult for me to put words onto paper, and I never understood why a summary was needed. It seemed redundant to me. My old classmates like to chide me about a paper I wrote in grammar school on the topic of metals. My teacher told me of the importance of summaries, so at the end of the paper I wrote: "In conclusion I would like to mention uranium." (Spangenburg 5)

But as you all know I was most infamous for my poor lecturing abilities. My students use to complain that I spent much of my class time attempting to light my pipe, and the rest of the time mumbling inaudibly to everyone accept those fortunate enough to have sat in the front row. I don't know how my students passed my classes. And yet, I did my best thinking while pacing and talking out loud. It helped me formulate my ideas.

RISA: You said that your parents were secularists. Was that a hindrance?

NIELS: No, on the contrary. Because my father was free of the shackles of religion, he was able to take up unpopular causes such as women's rights in a day when he had few allies. It freed me up as well, as I began to study philosophy and the great thinkers of the East like the Buddha and the Lao Tzu. It was because of these studies that I was able to see parallels between eastern philosophy and quantum physics. The ti and the t'ien were principles that were directly responsible for the formulation of the principle of complementarity.

You see, the whole history of western philosophy since the ancient Greeks has been synonymous with the philosophy of dualism, but the Buddhist and Taoist philosophy is neither dualistic nor monistic, but rather a unity of opposing forces which the Chinese refer to as the ti and the t'ien. It was precisely this that allowed me to understand that the wave and the particle were complementary. I believe it was my colleague sitting right here in your presence, who once proclaimed that if there is any religion that could cope with the modern scientific needs, it would be Buddhism. But I think that both Albert and I realize that Buddhism is not so much a religion as it is a philosophy.

By this time Albert is fighting to keep his eyes open.

ALBERT: Rashid, Risa, thank you so much for the pizza and the hospitality. I'm beginning to fade and I'm sure Niels is tired as well. It has been a long day. If you would kindly excuse us, I think I'd like to go home and turn in for the night.

RASHID: Oh, I hope we didn't keep you up too long. Thank you both for coming. It has been fascinating.

NIELS: Yes, it has. We'll have to talk again soon.

Niels and Albert get up to leave and Rashid and Risa see them to the door. On the way home to their apartments, plans are firmed up to meet for coffee on Friday. Over the next couple of days Albert and Niels keep themselves busy with academic work and they attend the all too frequent meetings and interviews around campus.

On Friday they meet for coffee and pastries as planned. Niels gets right to the topic at hand.

NIELS: Al, I have another experiment I'd like to show you that could answer some questions about what constitutes an observation.

Niels removes a device called a "holonet" from his briefcase. It is a device that projects a holographic image into the surrounding space producing a virtual bubble around Niels and Albert. Images appear as if they are on the inner wall of a sphere and as a result text and pictures have an oddly concave appearance. Although Albert has heard about the device, he didn't know that Niels had one and is quite amused by the technology. People who are not inside the hologram see nothing, so the rest of the people in the restaurant are undisturbed.

The experiment that Niels projects has been taken in part from an experiment described in Paul Davies book "About Time: Einstein's Unfinished Revolution" called the "quantum-eraser experiment." The experiment was carried out at the University at Berkeley.

> *A photon gun emits a photon and it is converted into two twin photons by a calcite crystal C. The photons are brought back together at a beam-splitter which transmits or reflects the photons with a 50/50 probability to detector (D1) or detector (D2). If the photon taking the left path is reflected, then the photon taking the right path will be transmitted and vice versa. Because the experimenter has no knowledge of which photon took which path, the photons go into a superposition of states resulting in both photons ending up in the same detector* (Davies, 169). (*Figure 5*)

ALBERT: I'm a little confused. Why do they go to the same detector if one is reflected and the other transmitted?

NIELS: Look at the position of the beam-splitter at position (B). Let's follow the route of a hypothetical photon. If a photon taking the left path gets

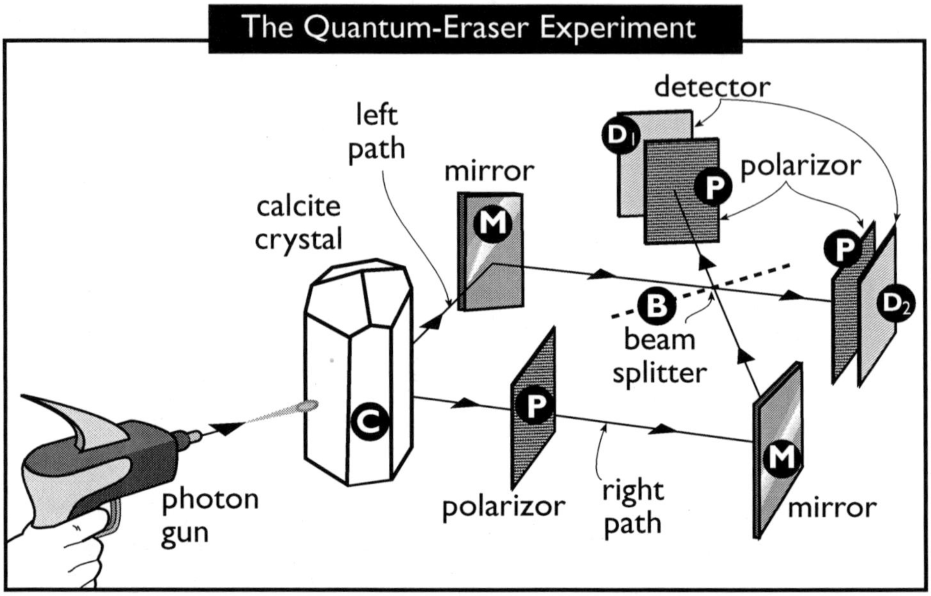

Figure 5.

reflected and bounces off the beam splitter and goes to detector (D1), then the photon taking the right path will be transmitted and will also go to detector (D1) and vice versa. If a photon taking the right path gets reflected and goes to detector (D2), then the photon taking the left path will be transmitted and also go to detector (D2) and vice versa. A 50/50 chance exists that the photon will be transmitted, and a 50/50 chance that it will be reflected. Since we don't know which photon took which path, the photon goes into a superposition of states and both photons end up in the same detector.

ALBERT: Oh yes, I see now. But what is the point of this experiment? It's sort of like the two-hole experiment.

NIELS: Well, we are about to add a very unique modification to the experiment. Take a look at what the experimenters did to modify the experiment.

Niels raises the holonet and projects the modified version of the experiment into the space before them.

In this new version of the experiment, a polarizer (P) is put into the right path that will change the photon's polarization by giving it a ninety degree twist. This measurement gives information about which path the photon took. Now the photons' superposition of states is destroyed and they act independently triggering both detectors simultaneously.

In the second part of this experiment, a polarizer is put in front of each detector which twists the polarization another ninety degrees and thereby erases the original information from the detector in the right path. As a result the original superposition of states is restored and both photons end up in the same detector once again despite the fact that they have already passed through the optical system at (B). (Davies, 17)

ALBERT: Ok, I'm a bit slow today. Can you walk me through this, Niels?

NIELS: Yes, sure, it is very confusing unless you spend some time thinking about it. In the original experimental set-up, the photons went into a superposition of states just like in the two-hole experiment. We have no knowledge of which photon took which path. As a result an interference pattern occurred at the beam splitter (B). The two probabilities caused the photon to merge and go to one detector or the other.

But in the above experimental set up, a polarizer is added to the right path which gives the photon a ninety-degree twist. Since a measurement occurs due to the polarizer, knowledge or potential knowledge is gained about which path the photon took. As a result of a measurement the photons' superposition of states is lost and the photons acts independently when they arrive at the beam splitter (B). The beam splitter (B) causes the merged photon to enter another superposition of states as it travels to the detectors and both detectors are triggered simultaneously.

Now here is the really weird part. If we then place a polarizer in front of both detectors which give the photons another ninety degree twist to reestablish their original polarization, we once again lose information as to which path the photon took when it left the source calcite crystal and now only one of the two detectors is triggered. This is really odd because the information appears to be erased retroactively, if you will, well after the photon has gone through the optical apparatus at the beam splitter.

Albert: Yes, yes! I think I understand. This experiment has two distinct features. First, we have the annoying backwards-in-time scenario again as in the two-hole experiment, but the experiment also raises the questions of what constitutes knowledge and what constitutes an observation?

We don't know, nor could we find out, what the original polarization of the photon was when it was measured and here is why. Although, Davies doesn't specify how the polarizers are constructed, I can say with some assuredness what type of apparatus was used in this experiment, because I've seen similar set-ups before. What he calls a polarizer is really two closely-spaced, twin polarizers that are each offset by 45 degrees, such that the first polarizer of the pair twists the photon 45 degrees and the second one twists it another 45 degrees. By doing it this way, one doesn't know the photon's original polarization because the first polarizer (set at 45 degrees) can't distinguish between a photon that is vertically polarized from one that is horizontally polarized. Therefore, the experimenter doesn't gain any real knowledge. After the photon leaves the polarizer, however, we do know that its polarization has been turned 90 degrees, but we don't know if it has vertical or horizontal polarization.

Niels: That's right. Of course a photon has no polarization until it is measured, but the concept is sound. This experiment demonstrates that just the mere possibility, or threat, of gaining knowledge is enough to change the outcome of the experiment.

Albert: Yes. And the second sets of polarizers, one set at each detector, work in the same way. Never do we know the actual polarization of the photon. So what constitutes information? No consciousness is raised in this experiment, so wouldn't that remove any possibility of an individual observer effect?

Niels: Al, the importance of this experiment is that the observer is for all practical purposes removed from the experiment. So, yes it would appear that a conscious observer is not necessary. This leaves us with the possibility that the universe at large is the ultimate observer. I'm just going to have to sleep on this one.

Two college students shyly approach Albert and Niels and sheepishly ask Albert for his autograph. They did not recognized Niels, so Albert introduces his friend and the students ask for his autograph as well. Albert loves the attention and engages the two in conversation.

ALBERT: How are you today?

STUDENT: Very well, thanks. We really admire the two of you, and we want to welcome you both to our campus. Everyone is really excited to have you here.

ALBERT: It's our pleasure. What are you studying?

STUDENT: I'm studying economics, and she, foreign languages.

The two students begin to feel uncomfortable for butting in and, not knowing what else to say, decide to leave the two scientists alone.

STUDENT: Well, I'm sure you're busy, so we should go. Thanks for the autograph.

ALBERT: Sure, it was nice meeting you both.

Niels is a bit more put out by the interruption and seems annoyed, immediately resuming their private conversation.

NIELS: Al, last night I started thinking about this teleology thing you were talking about, and I suddenly realized that there is a similarity between the indeterminate process of nuclear decay and something called the placebo effect.

ALBERT: Really! I thought I had convinced you that nuclear decay was a deterministic process.

NIELS: Well, I could see no flaw in your logic when it comes to the process of nuclear decay; however, it is a bit more difficult to imagine the deterministic model for the jump of the electron from one orbit to another around the nucleus of the atom. Both of these processes are similar; both are indeterminate. Your presentation has not convinced me of the determinacy of the quantum world, but I have not discarded your idea entirely, and it is possible that it might have implications for the placebo effect.

ALBERT: Very good, I'll take even the smallest of victories for now. So, how is the process of nuclear decay related to the placebo effect?

NIELS: Actually, the placebo effect could more appropriately be called the placebo response because the placebo which is an inert substance, such as a sugar pill or saline solution really isn't causing any kind of effect. Rather, the "effect" is really a response to an expectation that a subject or patient is getting a beneficial, or in some cases a detrimental substance, when in fact they are getting only a placebo. At least this is the current understanding.

Several things about the placebo response remind me of some of the things we encounter in quantum theory such as the observer effect and probability and chance. For example, one very curious thing is how groups of individuals in placebo studies seem always to maintain the same ratio of response to a placebo.

ALBERT: Is there some kind of personality trait or genetic predisposition for responding to a placebo?

NIELS: No, apparently not. Research has shown that this statistical phenomenon is remarkably consistent throughout time, and seems to be independent of one's culture, race, age, occupation, religion or psychological make-up. I read an archive article in *Health Magazine* that emphasized this fact:

> *Perhaps surprisingly, a person's personality does not predict whether he or she will show a placebo response. In fact, responding to a placebo in one situation does not mean that a person will show a placebo response in a different situation. (Turner, 1995)*

A long standing assumption suggests that a positive attitude and a strong desire to get well are essential elements among those who have been able to overcome their illnesses. This evidence is mostly anecdotal and does not account for those who have had an equally strong desire to live, yet succumb to their illnesses. The belief that we can overcome illness by means of a strong *desire* or *will* is tempting to embrace, since most of us have an emotional proclivity for personal control over our lives, but little evidence exists to support such a conclusion. Since a subject's response to the placebo effect in one double-blind experiment is not an indicator of response in subsequent experiments, and, since we can assume that the *will* or *intention* to get better is shared by most or all of the participants in a controlled placebo study, we must acknowledge that *intention* alone has a limited impact.

Al, this situation is not to minimize the effect that emotional states can have on the body. For example, fear or pain can trigger numerous neurochemicals responses in the body, such as the release of pain-reducing opioids, epinephrine, and corticosteroids, but these responses are too short in duration to account for the placebo response. In addition these emotional triggers are universal and would not account for the mere 30% efficacy rate. It appears that there are more basic processes at work which contribute to the surprising fact that regardless of one's culture, race, age, occupation, religion, or psychological make-up, the probability of response in a placebo study remains constant. However, many pharmaceutical companies remain unconvinced of these facts, and will often attempt to cull responders from a group in order to get the maximum benefit from their drug studies.

Al, suppose that a drug company is testing a drug called... oh I don't know, let's call it *Pharma*. Suppose they gather one thousand participants for their study. In an attempt to reduce the number of placebo responders some drug companies will conduct a preliminary study with a previously tested drug already on the market and a placebo. They do this in the mistaken belief that if they get rid of the placebo responders in the preliminary study, the drug they're testing in the main study will appear more efficacious. In most cases about one third of the group in the preliminary study will respond to the placebo as if it were the "real" drug. The researchers remove the placebo responders from the first group and keep those that did not respond to the placebo for the main study, believing that they have gotten rid of the weak-minded placebo responders. But, this effort is in vain. When the main study is conducted to test the *Pharma* drug with the new group, whose participants did not respond to the placebo in the first study, the researchers are miffed to find that approximately one-third of the remaining participants respond to the placebo. In other words, about two hundred and twenty people out of the remaining group of 660 will respond as if they had been given the real drug. Actually, a beneficial drug is supposed to do statistically better than the placebo in order to be efficacious, but normally the tested drug does only marginally better than the placebo. So it appears that not responding to a placebo in one study does not reduce the probability of responding to a placebo in a subsequent study.

Why is it that a person who doesn't respond to a placebo in one study will respond in a second or third study? And why is it similar to nuclear decay in which a percentage of the whole respond? You attributed this to different configurations of the universe. Each configuration is a new toss of the dice. Could this be what is going on here as well? Is our reality constantly changing depending upon the universe in which we live at that moment in time? It appears, Al, that we could figure out the half-life or more appropriately the, *third-life,* of a group of people in terms of their likelihood of responding to a placebo. Any group of people will respond to a placebo given enough time, or in this case given enough experiments. Just as in your idea of different configurations of the universe producing a new throw of the dice, so each experiment is a new throw of the dice for a group of individuals in a controlled experiment. In both instances, whether we are dealing with a group of people, as in the case of the placebo response or the entire mass of the nuclear element, the whole is determinate but the individual person or atom is subject to the laws of probability. At any time during an experiment all individuals within the group have an equal probability of responding or not responding. Those that don't respond in one experiment will still have the same chance of responding in the second experiment—one chance in three. This is a very odd situation.

ALBERT: Yes it is. So is belief of no importance at all?

NIELS: As I understand it, belief is very important in certain circumstances. In double-blind studies when the patient is uncertain whether or not they are getting the real drug, the belief system is in limbo because of the uncertainty, but in other cases belief is very important. Some researchers divide the belief factor into three separate categories: expectancy, conditioning, and meaning. I recently ran across an interesting experiment in Howard Brody's book, *The Placebo Response:*

Niels uses his device to project the article for Albert to read.

> *In 1970, Wolf's findings were taken a step further by Dr. Thomas Luparello and colleagues in the Department of Psychiatry at the Downstate Medical Center in Brooklyn, New York. Dr. Lulparello's team recruited twenty patients who had been diagnosed with asthma then exposed them to two drugs. One, isoproterenol, is commonly used in the treatment of asthma to expand the bronchial tubes of the lung, allowing air to flow where breathing had previously been obstructed. The other medication, carbachol, is a bronchoconstrictor and has exactly the opposite effect. Asthma would be expected to worsen when that substance is inhaled. After the subjects inhaled the drugs, the investigators measured air flow and lung volumes. Both drugs were given under two sets of conditions: one in which the subjects were told before- hand what the drug was; and the other in which the subjects were told beforehand that it was the other drug.*
>
> *Each drug worked better on average when the subjects expected its effects. That is, when told they were getting isoproterenol and were actually being given isoproterenol, the subjects showed more flow of air into the lungs than when they took isoproterenol, believing it was carbachol. When informed they would be taking carbachol and then actually got it, their lung flow worsened more intensely than when they were told they were taking isoproterenol but actually got carbachol.*
>
> *Of special interest here is what happened when four subjects given isoproterenol and five subjects given carbachol were told they were getting the other drug. The four subjects who got isoproterenol but thought it was carbachol showed an actual worsening of air flow in their lungs—consistent with the drug effect they expected, yet absolutely inconsistent with the effect produced by the chemical substance they inhaled. Five of the subjects who*

inhaled carbachol thinking it was isoproterenol showed an improvement in air flow—again demonstrating that the force of mental expectancy could overcome and reverse chemical effect. (Brody, 58)

NIELS: As you can see, Al, expectancy and belief are extremely important. And belief is not limited to taking pills or inhaling substances. The effect is seen in surgery as well. Let me call up another article I read. Ah, here it is. This is from biologist Bruce Lipton's book, *The Biology of Belief.* I'll project this out for you so you can read it.

Niels pulls out his holonet and projects the image the article.

A Baylor School of Medicine study, published in 2002 in the "New England Journal of Medicine" evaluated surgery for patients with severe, debilitating knee pain. [Mosley et al, 2002] The lead author of the study, Dr. Bruce Mosley, "knew" that knee surgery helped his patients: "All good surgeons know there is no placebo effect in surgery." But Moseley was trying to figure out which part of the surgery was giving his patients relief. The patients in the study were divided into three groups. Moseley shaved the damaged cartilage in the knee of one group. For another group, he flushed out the knee joint, removing material thought to be causing the inflammatory effect. Both of these constitute standard treatment for arthritic knees. The third group got "fake" surgery. The patient was sedated, Moseley made three incisions and then talked and acted just as he would have during a real surgery—he even splashed salt water to simulate the sound of the knee-washing procedure. After 40 minutes, Moseley sewed up the incisions as if he had done the surgery. All three groups were prescribed the same postoperative care, which included an exercise program.

The results were shocking. Yes, the groups who received surgery, as expected, improved. But the placebo group improved just as much as the other two groups! Despite the fact that there are 650,000 surgeries yearly for arthritic knees, at a cost of about $5,000 each, the results were clear to Moseley: "My skill as a surgeon had no benefit on these patients. The entire benefit of surgery for osteoarthritis of the knee was the placebo effect." (139)

NIELS: It turns out the typical one-third rate of response due to belief can be amplified. Instances have been recorded in which neither the physician nor the patient realize that they are getting a placebo. For example, in cases where mistakes have occurred such as in mislabeling drugs, both the doctor and the patients believe they are getting a beneficial drug. In these medical mix-ups it has been found that when both the physician and patients think they are getting the real drug, the placebo effect can double. The thing that is particularly worth noting is that unlike double-blind studies, in which a patient or the doctor might defer any expectations of the efficacy of the drug due to the fact that they aren't sure if the patient is actually getting the real thing, this is a case of total ignorance. Both the doctor and the patient have full expectations that the drug will be efficacious. We saw earlier how determinism and probability plays a role in the placebo effect when we are dealing with the typical double-blind studies. Now, we see how the observer effect comes into play as well. It turns out that belief is a very powerful placebo event just as an observation is an important event in the two-hole experiment.

ALBERT: Let me try to understand this. In a placebo study where the patient is not sure if they are getting a beneficial drug, about one-third of any group tends to get better because they are part of the experimental set-up in which determinism says that one-third will respond regardless of the make-up of the group. But now you are saying that belief is important; not only the belief of the patient, but the attending practitioner's belief as well. Some would make the charge that the doctor gives subtle signals to the patient when he or she knows that the patient is getting a real drug, but if the doctor is also fooled, as in your example when there is a mislabeling of drugs, then that would rule out the possibility that the doctor is inadvertently giving some kind of subtle signal to the patient. So, in that case, you are asking me to believe that the doctor's belief alone is enough to change the situation; the doctor's belief, all by itself is capable of healing a patient.

NIELS: Yes.

ALBERT: Niels, you compared the belief of the doctor as an important factor in the patient's healing process and you compared this to the two-hole experiment in which knowledge or belief collapses the wave function, but you and I both agree that one cannot *will* the particle to go through one hole or the other. It is nature's choice. Now you seem to be saying that a doctor can *will* the patient to get better.

NIELS: Well, you might think that I'm contradicting myself; however, there is a difference between our *will* and our *belief*. When we *will* something, it is our conscious mind, what Freud called the "ego", the temporal aspect of the mind that is wishing for a particular out-come. I demonstrated previously that will or intention has little impact on getting better. But beliefs are more deeply seated. Beliefs reside in the subconscious, the "id." The conscious mind knows that what we wish is not necessarily rooted in reality, but the same cannot be said of a deep- seated belief. The subconscious mind cannot distinguish between a belief and objective reality. Belief and reality are for all practical purposes the same. In fact, many say that consciousness is reality. Beliefs reside in a wave form—a superposition of states. When we become consciously aware of our beliefs, it is similar to performing a particle experiment in that the observation by our aware conscious collapses the wave function in the mind and in the universe at large.

Al, remember when we were reviewing several commonly performed two-hole experimental set-ups? Each had a predictable outcome under the laws of quantum mechanics. The wave/particle duality expressed itself depending upon whether or not it was observed, but then, we ran across the inexplicable quantum-eraser experiment. The patient, who was supposedly having his knee operated on, believed that he was getting the real surgery, but it was a fake surgery. In the quantum-eraser experiment, nature believed information was being extracted about the polarization angle of the photon, yet no meaningful information was actually gained, and yet it influenced the outcome of the experiment. Could it be that the two situations are equivalent? Look at this:

Niels uses his device to project a holographic scene.

Fabrizio Beneditti at the University of Turin in Italy and his colleagues induced pain in subjects, and then controlled the pain with morphine. After a few days, the regiment of morphine was replaced by a placebo, a saline solution, without the patient knowing about it. Nevertheless, the pain was reduced just as if subjects were getting the morphine. One could say that this was due to the patient's expectations of getting the "real" drug. But here is the twist. Once again, without the patient knowing it, the researcher added Naloxone to the saline solution. Naloxone blocks the effects of morphine. This should have had no effect since the patient was not getting morphine, only a placebo, and yet the patients reported that the pain had returned. (Brooks).

NIELS: The patient had no expectation that a morphine blocker was being added to the saline solution, nor should there have been any effect to adding a morphine blocker to a saline solution, so just who was it that was being fooled here? Was nature herself being fooled? Al, could we draw a comparison between the quantum-eraser experiment and the placebo experiment?

ALBERT: Be my guest.

NIELS: We can equate the patient in the placebo experiment with the set-up of the quantum-eraser experiment. When the doctor injects the patient with the placebo, he is aware that he has changed the information in the experiment; however, the patient still believes she is getting the morphine. Because of the doctor's knowledge, there is a collapse of the superposition of states. As chance would have it, the patient in this particular study got the benefit of the placebo as if it were the morphine injection. She was one of the lucky one-third responders. The doctor's belief collapsed the superposition of states into a single reality. Even though the patient gains no knowledge, it appears that nature has gained knowledge and as a result the placebo takes the place of the morphine. Now, when the doctor injects the Naloxone the experiment is changed again. Just as the information was erased when the polarizers were placed in front of the detectors in the quantum eraser experiment, so too, was the information erased when the patient received the Naloxone. No information was known to the patient, and no information was known to the researchers in the quantum-eraser experiment, yet the phantom information was erased in both the quantum-eraser experiment and in the placebo experiment. So, to make this perfectly clear: Just as the experimenters in the quantum-eraser experiment did not believe that adding polarizers in front of the detectors would erase the information, so the doctor in the placebo experiment did not believe that the Naloxone would block the effects of the morphine. Yet, the information was erased in both cases. I think what is happening here is that even though information was extracted, meaning was not. Since the information had not established a history, since it raised no consciousness in either case, it had no meaning. Al, I have a suspicion that meaning and information are both essential aspects for understanding these experiments.

ALBERT: Hmmm. I think you are on to something there. I've been thinking along similar lines. David Bohm and I talked about meaning in some depth after the war. I'd like to tell you about those discussions when we have more time.

NIELS: I'm wondering, Al, if it is possible to explain some aspects of these experiments in terms of the unusual properties of water and its interaction with belief or meaning.

ALBERT: How so?

NIELS: Well, as you know the chemistry of life is impossible without water. In addition water has many bulk properties which make life possible. For example, unlike other substances, water becomes less dense as it cools which makes ice float and prevents the oceans from freezing from the bottom up. Water also resists warming and cooling, regulating the climate by making it more stable and, thus, allowing species time to evolve. Water's boiling point is 150 degrees centigrade above comparable liquids and it has a very high surface tension. But it is water's structure—its quantum effects that are of real interest here.

We tend to think of water as having a fairly stable structure, but the hydrogen bond between water molecules is anything but stable. Water molecules only attach to their neighbors for microseconds, as new bonds are constantly being broken and reformed with other molecules of water.

This chaotic behavior is due to the weak electrostatic bonds between the hydrogen atoms. These bonds are ten times weaker than the covalent bonds between the oxygen atom and the hydrogen atom within each water molecule. The zero-point energy of space provides the energy and vibrations which breaks and recombines the individual neighboring molecules of water. It turns out that the quantum effects of water are responsible for the very function of proteins, and DNA.

Al, I think it was in the late 1990s and early 2000s that the United States government as well as independent companies launched the human genome project: a project designed to catalogue thousands of gene sequences that make up our DNA. Biologists were expecting to catalogue around 100,000 genes because there are about 100,000 proteins. It was assumed that it took one gene to encode one protein. But scientists were shocked to find only around 15,000 genes. It turns out that proteins perform various functions according to their geometry—their physical shapes. One protein can fold into as many as a thousand distinct shapes, and as a result one protein can carry on a multitude of functions as it transforms from one shape to another. This shape-shifting and folding would take years to happen under the laws of classical physics; however, under the laws of quantum physics, namely quantum tunneling, transformations can happen in microseconds. Water plays a definitive role in this procedure. It was found that protons must move through the protein to affect the change of shape. Water, trapped inside of the protein provides the source for these protons—the nucleus of the hydrogen atom. When the proton is hit by a photon of light, a kind of photosynthesis takes place providing the energy for the protons movement through the protein, which in turn changes the shape of the protein.

A similar process takes place in the DNA as well. It has recently been discovered that as water molecules get near a DNA molecule, the weak hydrogen bonds are disrupted and they form a template around the DNA molecule and through electrostatic forces, signal the approaching protein of the positioning of the DNA bases. The proteins take this information and align themselves for the best "fit" and establish contact with the DNA molecule.

ALBERT: So what does this have to do with the morphine placebo experiment?

Niels: Well, many in the new-age community have long suggested that water has memory. This is the basis of the pseudo-science of homeopathy which I'll get to in a moment. I believe that it is possible that water in the body, which is unobserved, goes into a superposition of state. This superposition could conceivably form a template for the formation of a morphine molecule. This would provide an explanation as to why the Naloxone was able to block the saline solution that had literally turned into the morphine. Chemical templates have been used when creating crystals in the laboratory and in cloud seeding. Silver iodide provides a template for the formation of crystalline ice particles. So, chemical templates are not unheard of in science.

Albert: Interesting.

Several college students recognize Albert and Niels and stop to get autographs, but this time Albert doesn't try to engage them in conversation as he knows it irritates Niels.

Albert leans back in his chair and makes a contemplative sigh, as if he wants to say something.

Niels: What are you thinking?

Albert: I'm just thinking about what it was like when I was their age. We had it all and we didn't even realize it. Do you ever think about those days and think about how great we had it?

Niels seemed surprised at Albert's question and had to shift gears as he was still thinking about the placebo response.

Niels: Ya, Often I do.

Albert: Tell me about it.

Niels: I believe it was about 1911. I had just defended my thesis and dedicated it to my father who was everything to me. Before his death, my father secured a position for me to work at the Cavendish laboratory with the great J.J. Thompson in Cambridge, England. He had won the Nobel Prize in1906 for his discovery that the atom was composed of electrons. He called them corpuscles I believe. He thought that the electrons were interspersed throughout the atom like raisins in a cake.

It's hard to describe the feeling I had when I got to Cambridge. I was walking the grounds where Isaac Newton and James Clerk Maxwell once walked, and if that wasn't enough, I was about to meet and work with a Nobel Prize winner in physics.

I rented an apartment with a fireplace, joined a soccer club, and took nature walks contemplating what my first meeting with Thompson would be like. I wondered just what I would say, and what he would say when I handed him my dissertation which had to do with the way electrons behave in metals. Well, that meeting finally came and Thompson was very personable. I gave him my paper and he promised to read it and get back to me to discuss it. I spent the next weeks in anticipation of hearing from him. I remember writing letters to my fiancé Margrethe and my brother Harold, bragging about what a great life I had at Cambridge. But days turned into weeks with no word from Thompson. Finally, I got up the courage and contacted him again. Turned out he hadn't even picked up my paper from the place I laid it on his desk. He made more promises to read my paper, but it never happened. I was crushed. Things went from bad to worse. I just wasn't fitting in with the group I hung around with in Cambridge. I was soft spoken and my English wasn't very good, and that didn't help matters. When I did break through and make a suggestion, my colleagues would brush it off, as if it was a silly idea, and then turn around and say basically the same thing to the rousing approval of their peers. I wasn't part of their club. I began to feel very alone, and so I made a decision to go to Manchester to work with Ernest Rutherford. That turned out to be the best decision of my life.

Rutherford was a brash and boisterous experimental physicist from New Zealand who had already accomplished a great deal by the time I arrived in Manchester. He had little use for theoretical physicists like me, yet, ironically, we hit it off immediately, thanks in part to the fact that I was a footballer and he respected that.

Rutherford had gone to the Cavendish laboratories as a young man to study with Thompson, but he had much more to show for it than I did. When he was working with Uranium, he noticed that the element spontaneously emitted two distinct particles, the Alpha particle, which we now know to be the nucleus of the helium atom, and the beta particle which was later found to be electrons emitted from the nucleus. He published a paper on the subject in 1899 I believe.

After Cambridge, Rutherford went to Canada to accept an appointment as a professor of physics at McGill University. It was there while working with Fredrick Soddy that he made a discovery that would win him the Noble Prize in chemistry. The two experimentalists noticed that the radioactive element, thorium, radiated a gas, which when analyzed, turned out to be the inert element argon. This was indeed a stunning revelation because it clearly demonstrated that one element can spontaneously transmute into another element—radioactive decay.

In 1908 he returned to England to accept the position of experimental physics at the University of Manchester, where he collaborated with Hans Geiger. Here the two men set up an experiment in which they fired alpha particles at pieces of gold foil. You'll probably remember, Al, what happened next. Some of the particles passed through the foil unimpeded, while other particles were diffracted at acute angles. Rutherford pondered the situation, and in one of those inspirational moments that you know so well, he realized that J.J. Thompson's model of the atom could not be correct. From his experiments, Rutherford believed that the atom must have a densely packed nucleus with light, negatively charged particles orbiting the nucleus. This seemed to me a very big leap of intuition, but it explained why some alpha particles passed right through and other particles that struck the hard nucleus bounced back. His idea was getting closer to the model of the atom we know today, but we were not yet there.

So, now, as destiny would have it, I was going to work for the "Newton of Atomic physics." So much for the idyllic setting and charm of Cambridge; I was now working in the sooty, filthy, industrial city of Manchester, in a dark and dingy laboratory where we had to duck under heating ducts and step over electrical cords and water pipes. But no one cared. We were doing cutting edge physics and every day when we went into the laboratory we expected something great to happen. Everyone knew that Rutherford was in charge, but there was always a free exchange of ideas and, unlike Thompson, Rutherford was willing to listen to our ideas, and that was the reason that the lab was so productive. This is where everyone wanted to be. We were a real team. Everyone contributed.

ALBERT: What did you think about Rutherford's model of the atom at that time?

NIELS: That is where my training in theoretical physics really paid off, because I immediately saw a couple of problems with Rutherford's model. For one thing, if the electron orbits around the nucleus as Rutherford proposed, then the electron would necessarily produce electromagnetic radiation as it accelerated around the atom, yet most matter does not radiate light or other forms of radiation. Also, if the electron moved around the nucleus, then according to Newtonian mechanics, it should very quickly lose energy and spiral into the nucleus, but this obviously didn't happen either.

During that period of time I had been interested in the emerging field of quantum mechanics, especially your work and Max Planck's work. Planck showed that light radiation emitted from a black body came in discrete units, and you made the intuitive leap that light itself must be quantatized. That gave me the impetuses to formulate a non-mechanical description of the atom. I really believed I was on the right track, and I couldn't wait to think deeply about this idea, but I was also thinking deeply about Margrethe back in Copenhagen. I left England to go home and make her my wife.

ALBERT: Did you have a big wedding?

NIELS: No, we just had a civil ceremony, but we had a great big party afterwards. After the party, Margrethe and I took a ferry to England for our honeymoon, and during that trip I finished my paper on the absorption of Alpha particles.

ALBERT: How romantic.

NIELS: We continued on to Manchester and got together with Ernest and his wife. We all became very good friends and had a great time together. I gave Ernest my paper and then we headed back to settle in Copenhagen.

ALBERT: Is that when you continued your work on a non-mechanical description of the atom?

NIELS: Yes, I was thinking about this all the while and I could barely wait to get back to it. If you want to know the truth, I was deathly afraid that you would beat me to the punch. It seemed that the idea I was working on was very similar to what Planck and you were doing, so I thought that time was of the essence. I also remembered that when you were working on special relativity you were able to solve a couple of mysteries with one theory, and I began to think that I could do the same with my idea. At that time, no one knew why each element had different spectral lines. So, I began to mathematically calculate the energies involved in the jump of the electron from one orbit to the next in the hydrogen atom, which was the simplest to work on, since it has only one electron. When I discovered that the radius of the electron's orbit was directly related to Planck's constant, I knew for sure I was on the right track. If the electron absorbed and emitted energy in whole units, in Planck units, then it became clear that there could be no in-between orbits. This is why the color spectrum doesn't fade gradually from one color to the next. The spectrum's colors are in discrete bands. It turned out that the frequency of light emitted when an electron goes from one orbit to the next was in multiples of Planck's constant. That also explained why the electron did not spiral into the nucleus. An electron must always have one Planck unit of energy. Just as you had the happiest moment of your life when you realized that a person in free-fall doesn't experience the effect of gravity, I had my happiest moment. I knew I had hit pay dirt. Now the only thing I worried about was getting the thing published before anyone else did, but I was thinking that someone could well be you. I struggled to write a paper, which as you know has never been my strong point. After many attempts, I finally hammered something out and showed it to Rutherford, but he didn't think it was ready for publication for a couple of reasons: First, he thought it was way too wordy. Can you imagine that? The other problem was that ole Ernest was still in the world of classical physics. In particular, he wondered how the electron knew when to jump from one orbit to another. This was a question to be answered by quantum theory. It comes down to pure chance and probability, but Rutherford, you, and others were not yet ready to accept that. Despite these flaws, he thought that the paper had great merit and he said I should go ahead and publish. Rutherford was right on one count. It was too long for publication, and he suggested I cut it down to size, but I was too stubborn, so in 1913 I was able to get it published in the form of a trilogy, and in 1922, as you know, I received the Nobel Prize for that work.

So to answer your question, I do think about those days a lot, and I would have to say those years were the most exciting times of my life. We humans are curious creatures who strive for novelty and the pursuit of the secrets of nature is our highest calling. The idea that we could do experiments and then through reason actually discover something new and formulate our findings into mathematical equations; that's nothing but pure joy. Unlike you, Al, I only experienced this feeling one time in my life. You had such experiences time and time again. What was that like for you? Tell me what led up to your achievements.

ALBERT: Yesterday when we were with Risa and Rashid, you heard about my early childhood. I guess my first remembrance of that certain sense of mystery came when I was sick in bed and my father gave me a compass. I was simply mesmerized, and that one simple act by my father instilled a sense of wonder in me that lasted my entire life.

NIELS: Where did you go to school?

ALBERT: When I turned six, my parents sent me to a Catholic school, just because it happened to be in our neighborhood. Of the seventy students or so, I was the only Jew. Even though I didn't have a religious upbringing I became really interested in my Catholic studies courses, and I did so well, that I helped out my classmates.

Then when I was around nine years old, I attended an institute that emphasized math and science as well as Latin and Greek. They also had religious classes for Jewish students which I soaked up like a sponge. Oddly, I developed a passion for Judaism as well, and I began to participate in Jewish rituals and I celebrated Jewish holidays to the complete bewilderment of my parents. It was just fascinating to me for some reason.

NIELS: Did you have trouble with math when you were young or is that just popular myth?

ALBERT: People love to say that I flunked math when I was young. I think it makes them feel as if they have some hope of greatness despite their deficiencies. The fact is, that I was very good in math, and, by the time I was twelve, I was about two grade levels ahead of the rest of my classmates. While they were working with arithmetic, I was solving geometry and algebra equations. My parents saw my aptitude in this area, so they bought me geometry and algebra textbooks so that I could study over the summer vacation. I was totally consumed by mathematics at that time, and by the age of fifteen, I had mastered differential calculus. While other kids were out playing, I was engrossed in my newfound love of math. It was foreign languages that I was so poor at. I nearly flunked out of Latin and Greek. As you know language was never my strong suit. I must be right brained.

I also owe much to my parent's generosity which indirectly helped me with my studies. You see, in those days it wasn't unusual to help out struggling college students, so my parents took in a needy medical student named, Max Talmud, to share meals with our family. Max brought me science books written by Aaron Bernstein, called *People's Books on Natural Science.* Bernstein published over twenty of these books which emphasized the interrelationship between biology and physics and those books included detailed scientific experiments (Isaacson, 18). It was while reading those books that I knew which direction my life would take. I was totally infatuated with natural science. Many of my ideas about relativity came directly from those books.

NIELS: How so?

ALBERT: Bernstein often asked his readers to imagine real life scenarios demonstrating the principles of relativity. What would it be like to be on a train and observe a bullet fired through the window? What would happen if it were a beam of light instead of a bullet? I remember clearly that in one statement Bernstein made reference to the fact that, since electromagnetic waves propagate at the same speed, it holds that the speed of light must be a general law of nature (19). This idea is the very cornerstone of my special relativity theory. It is amazing just how close Bernstein was to discovering that the speed of light is a constant.

NIELS: You said that Bernstein's books tried to unite various disciplines of science into one. Did that help you formulate relativity as well?

ALBERT: Yes, Bernstein's tendency to see nature as an undivided whole influenced me to always look for the simplest and most elegant designs in nature. When I saw two apparently diverse explanations for similar phenomena, I looked for a synthesis of the two. When I was working on special relativity, I became aware that space and time were subjective concepts that had no validity in and of themselves, so I unified them into the singular objective concept of space/time. I saw the inconsistency of not including light into the Newtonian/Maxwellian frame of relative motion, and so, I included light into my theory. I found the common denominator of matter and energy and united them into single formulae. I called my new theory "special relativity" because it didn't include accelerated motion and that bothered me greatly, so I immediately went to work in an attempt to generalize the theory.

NIELS: If there is one thing we could say about you, besides being a determinist, it is that you are the great synthesizer and it wasn't limited to classical physics; you did the same thing in the quantum realm. When Planck discovered that the radiation emissions from a blackbody could be mathematically understood if the emissions were parceled out into packets of energy, you made the intuitive leap that light itself must be quantatized, and when Satyendra Nath Bose proposed that photons having the same attributes could be treated as if they were just one particle (since they are indistinguishable) you, realizing that mass and energy were equivalent, extended this theorem to include all atomic particles. You've made a career out of holistic inclusion.

ALBERT: I got a lot of mileage out of throwing everything into one big pot. I owe so much to Max Talmud because he really shaped my education at an early age. I remember working with some of the geometry books he brought me. After a time, I became so good at proving Euclid's theorems, that I had outstripped Max's mathematical abilities. He could no longer help me with my proofs, so he introduced me to philosophy, which I consumed with the same zeal as I did the math books.

By the age of thirteen, I was heavily into Kant, Hume, and Ernst Mach. Their philosophy helped me reinforce my general beliefs about nature. When I read Kant's work, I already had an understanding of Greek rationalism and the idea of *priori* knowledge—the idea that the universe can be understood through pure reason. Kant helped me understand that, though we can never know nature directly, we can still draw consistent analogies between appearances and reality. David Hume took Kant's brand of rationality to a whole new level. His brand of rationalism is what we now call logical positivism, which asserts that science can only address those things that can be measured or directly observed. These ideas helped me when I was working on special relativity. In those day it was believed that if light was a wave, then it needed a medium in which to propagate. In addition, if light was a relative concept, then it must move relative to something else. The ether fulfilled both conditions. It provided a medium as well as a reference point in space. But experiment after experiment failed to detect the ether. I believed that there was a simple explanation why the ether hadn't been detected. It didn't exist and it wasn't needed—not as a medium for light to pass through and not for a reference point. I discovered that light was a constant, and this created a couple of other problems. If the speed of light is the same whether you are speeding toward the light or speeding away from the light, then something has to give. That "something" was space and time. If the speed of light is a constant, then space and time must be relative. One of my greatest revelations was that distance and time are malleable. So, as it turned out, neither the ether, space, nor time could be measured objectively. The discovery of relativity was a reflection of Hume's philosophy of logical positivism.

NIELS: I remember that when Heisenberg was unable to solve the problems involved with his principle of indeterminism, he remembered what you had said about only paying attention to that which you can measure or directly observe. He gave you credit in helping solve the riddle of indeterminacy.

ALBERT. Yes, I remember that, but it was my friend Max Born who helped him with the mathematics.

NIELS: What about Ernst Mach then?

ALBERT: Mach was a positivist. Mach said that Newton's "absolute space" and "absolute time" are entities with no observational validity. He suggested that motion must be judged relative to other objects in the universe, rather than against a background of space. Mach reasoned that there is no such thing as a point in space, and therefore, space can have no intrinsic reality. And since time is related to motion, and motion can only be judged relative to other objects, time itself must be a relative concept.

NIELS: He was close to discovering special relativity too, wasn't he?

ALBERT: Yes.

NIELS: At this time in your life you were consuming books on science, math, and philosophy. What happened to your religious studies?

ALBERT: As I gained more knowledge in these disciplines I came to the realization that the stories of the Bible couldn't be true. I understood that religion had a debilitation choke hold on society, and I rebelled against every type of social, governmental, and religious dogma to the point where my teachers told my parents that my mere presence in class was a disruption.

NIELS: I went through the same thing with religion. People were always trying to force God on me.

ALBERT: Yes, I remember when I went to America in the 1930s. By that time my fame had spread around the world and I was mobbed by reporters. The one question I could be certain they would ask is: "Do you believe in God?" To just say no, always brought on more questions, so I worked on various ways to dance around the subject until they heard what they wanted to hear.

NIELS: How long were you in Germany?

ALBERT: We moved around a lot because of my dad's business ventures. He had a featherbed company in Ulm, but when that failed we moved to Munich where he joined his brother Jacob, who was an engineer, in a gas and electric business venture. At that time electrification was coming to Europe and it seemed like a good business opportunity. The company was quite successful for a time. They had over two hundred employees and this start-up company provided lighting for over ten thousand people in the suburbs of Munich. My dad and uncle built their own dynamos using giant gasoline engines to turn the turbines. For a while, it looked as if they might make a fortune, but as luck would have it, they lost some important contracts to other companies. The business failed and they decided to move the operation to Milan, Italy, but when that didn't work out, they moved again to Pavia where the opportunities were better. I was left behind in Munich in the care of a relative. My parents valued education above all, and they thought that I should finish out my schooling in Munich. I was devastated. This was the lowest point in my young life. The schools there were run like military camps and learning was by rote, without any real understanding of what we were expected to learn. There was no sense of discovery at all, and the teachers were mainly concerned with discipline. I really detested school, and my love of learning was completely squelched. I became a disciplinary problem. I was the kid with the big smirk who sat in the back of the class, blurting out the one-liners that sent the entire class into raucous laughter, and when the teacher turned his back, I would launch paper airplanes or drill someone with a spit wad.

NIELS: Ah, those timeless antics. What came of all that?

ALBERT: Well, they hadn't beaten the sense of adventure out of me, so I hatched an ingenious plan to free myself from this repressiveness and join my parents in Italy. I arranged for a doctor, who was a family friend, to write a letter to the school saying that I was suffering from a condition of nervous exhaustion, a fact that was not far from the truth. The scheme worked, I think in part because the school was happy to get rid of their disruptive and belligerent student. I had another reason to get out of Germany. I knew that if I stayed there until graduation I would be drafted into the army. That was a prospect I couldn't bear.

NIELS: Did your parents know about your scheme?

ALBERT: No, I packed my bags, bought a train ticket, and got out of Germany at the earliest opportunity. I don't think anyone can imagine the joy I felt as I rode that train out of Germany and into Italy. I moved in with my parents, joined my father and uncle in their electrification business, and promised them that I would study on my own so that I could pass the requirements to enter the Zurich Polytechnic Institute the next fall.

NIELS: What was that like, working at the business?

ALBERT: I enjoyed it and my dad and uncle were thrilled to have me there. Sometimes they would be struggling with engineering problems for weeks on end, and I would jump in and in a matter of minutes figure out how to do it. My father and uncle realized what an asset I was to the business and they wanted me to stay on, but it wasn't something that I would have wanted to do my whole life. I will say this though, it was an invaluable experience and I learned a lot about the practical aspects of working with coils, magnets, and electrical generators.

NIELS: Hey, Albert, look at the time.

ALBERT: Wow, we've been sitting here for almost two and a half hours. If we stay any longer, we'll have to order lunch.

The two get up to leave and make plans to get together again to continue discussions on the meaning of reality.

Over the next days both men work diligently in their offices and prepare for the experimental work that Princeton has assigned them to do. Most of the technical aspects of the upcoming experiments have been relegated to the engineers at NASA, but it is up to Albert, Niels, and John, to make sense of the data. In addition, both Albert and Niels must attend to the many requisite interviews and public appearances their fame has necessitated. The next opportunity the two men have for discussion occurs on a warm spring day. After lunch, and a leisurely walk around campus, Albert stops by Niels' office in hopes that he is not busy. Niels hears the knock on the door.

NIELS: Come in.

ALBERT: Are you busy?

NIELS: No, not right now, sit down and chat. I don't have any appointments until 3:00 this afternoon.

ALBERT: Last time we were together you were talking about water having consciousness. You talked about water forming a template to imitate the effects of certain drugs and that you believe water was able to orchestrate the synthesis of proteins in the body. Then you brought up the subject of homeopathy. What's that all about? Is there something to the homeopathy thing or have you gone soft on me?

NIELS: I have been thinking about homeopathy and how it relates to this whole discussion. Homeopathy is an alternative medicine that has been around for centuries, and its perceived efficacy has risen and fallen from one generation to the next. In recent years homeopathy has fallen under the scrutiny of the scientific process. Just when it appeared that homeopathic claims had been substantiated by a few scientific studies, follow-up studies failed to find reproducible evidence for its efficacy. Homeopathic advocates cannot clearly and unequivocally reproduce the results in a reliable manner.

ALBERT: So, what do you think is going on specifically?

NIELS: Well, it's very interesting how the process of homeopathic preparations proceed. Typically, a homeopathic practitioner begins with a toxic substance and adds alcohol. The tincture is then poured into a graduated beaker to, let's say the one-tenth mark, and water is added until the beaker is full and then the mixture is shaken vigorously. The new diluted substance is poured into a second graduated beaker to the one-tenth mark, water is added to fill, and it is then shaken. The substance has now been diluted one-hundred-fold. This procedure is repeated many times until the substance is so diluted that it may not have a single molecule of the original toxic substance. Homeopathic practitioners believe that the diluted substance still retains the molecular memory of the original substance without its harmful attributes. It is then consumed with the promise of healthful effects. In light of what we have learned about the placebo response, and the quantum effects of water, I believe it is quite possible that there might well be something to the homeopathic claims.

ALBERT: I'm kind of confused. You have tried to draw comparisons between the quantum- eraser experiment and the placebo affect and now homeopathy? I liked the idea that you have retained a somewhat deterministic approach to these things, but could you clarify this a bit more?

NIELS: Well, I did agree that your deterministic description of nuclear-decay seemed to be related to what occurs in double-blind studies in the placebo effect, but remember, I also said this did not account for the effects *belief* seems to have in these experiments. Remember that the quantum enigma is a two-tiered problem of probability and the observer effect. I equated the situation of unobserved particles with the concept of "belief" in relation to the placebo effect. Both of us seem to be uncomfortable with the anthropomorphic notion that we determine reality with observation, and as we saw in the quantum-eraser experiment, the observer had for all practical purposes been eliminated. Observation in the form of belief had also been eliminated in the morphine experiment. Neither the patient nor the doctor believed that the Naloxone would have any effect on the morphine and yet it did. It appeared as if nature was the observer. The situation with homeopathy is very similar. No trace of the original substance exists in the final dilution of that substance, and yet, water itself seems to retain a *belief* or *memory* of the original substance. This effect appears to be independent of the practitioner or the patient being treated, and yet, it appears to have a curative affect. This suggests that nature can act as an observer. Nature has consciousness.

ALBERT: Very interesting indeed. You haven't given up the idea of the observer effect, but at least you are now admitting that consciousness is objective. That is a huge step, Niels. I commend you. But it is still difficult for me to understand how the consciousness we have in our heads, which I assume is a product of the activity of the neurons in our brains, is related to this universal consciousness of which you speak. But more to the point, it is hard for me to believe that a belief, whether it originates from inside our brains or from nature herself can affect matter. How does a belief cause a physical change?

NIELS: Al, a few days back you asked me how a non-physical mind could affect non-physical matter. Perhaps you thought that I was reverting to Descartes' dualistic notions. As you know, Descartes was a mechanistic determinist, like you, who believed in the clock-work universe. He thought that everything could be predicted if the initial causes were known. Both chaos theory and quantum theory have dispelled that notion completely. He was also a behaviorist in that he thought even human actions were attributable to simple cause and effect. But he knew that this kind of thinking would get him in trouble with the church, so he threw out a bone. The idea that we are mere automatons, subject to causal influences and not responsible for our actions, was repugnant to the church. If we have no free will how, can we make moral decisions? In our times this argument is framed in a familiar refrain: How can we be responsible for our actions if we are a product of deterministic genes? How can we choose between good and evil?

His famous statement: "I think, therefore I am," was really a statement about free will. If we could break this down into a Greek Syllogism, his thought process might have gone something like this: The universe is deterministic and everything can be explained as a result of a former cause. I can think any thought that I please; therefore, my thoughts are not a result of a former cause so I must have free will. Or another way to put it might be something like this: I have free will, because I can think any thought I please. All matter is subject to the deterministic laws of the universe; therefore, mind must be independent and distinct from matter. This idea saved him from the dictum of the church. His main premise that we can think any thought we please is fallacious, and that was always the problem with the deductive method, but I'll leave that argument for another day.

At any rate, he did not believe that the mind could influence matter, nor was it likely he thought that mind was something ethereal, though we will never know that for sure. His was more a philosophy called *epiphenomenalism*. He believed that mind is an emergent property of the brain, but he didn't think that the mind could affect the brain. This idea has been called into question in recent times as well.

ALBERT: Let's hear it.

NIELS: Al, you remember Henry Stapp don't you?

ALBERT: Yes, I knew him. He was a friend of Heisenberg.

NIELS: Henry thinks that the brain is a quantum mechanical structure. The brain contains about 100 billion neurons, and each of these neurons forms thousands of synaptic connections with other neurons. Genes are responsible for the main pathways and some of the initial connections, but the enduring connections are a result of external input: stimuli from the environment. At each connection lies the narrow synapse only about one nanometer across. Electrical signals are transferred through the synapse in the form of charged particles or ions. These are positively charged sodium and potassium ions, or negatively charged chloride ions (Schwartz, 104). The key here, Al, is that the ion channel is so narrow that there is a great amount of uncertainty. Since the ion's position is extremely limited as it passes through the channel, then, according to Heisenberg's principle of uncertainty, its velocity must be great (358). As the ion emerges into the vesicle which houses the *trigger release sites* for the neurotransmitters, the probability of a hit or a miss goes into a superposition of states of "release" or "not release." If the chemical is released, the signal continues; if not, it dies there and no message is passed on. No action is taken. The brain is constantly producing trillions of quantum states, and it is Stapp's belief that when the independent mind focuses attention on a particular question, it collapses the superposition into one definitive thought. This thought can be held in consciousness by repeatedly asking the question. William James called this the stream of consciousness. Today's scientists call this the quantum *Zeno effect.*

ALBERT: I don't think I've ever heard of this quantum Zeno effect.

NIELS: The quantum Zeno effect was named after the Greek philosopher Zeno who thought that motion was discontinuous. The quantum Zeno effect describes what happens when certain quantum attributes exist in a superposition of states. An observation will collapse the superposition into one particular state, and continual observations will hold it in that state indefinitely.

Jeffrey M. Schwartz demonstrates this process with the following experiment.

> *An ammonia molecule consists of a single atom of nitrogen and three atoms of hydrogen. The arrangement of the four atoms shifts over time because all the atoms are in motion. Let's say that at first the nitrogen atom sits atop the three hydrogens, like an egg nestled on a tripod. (The nitrogen atom has only two options, to be above or below the trio. It cannot be in between.) The wave function that describes the position of the nitrogen is almost entirely concentrated in this configuration; that is, the probability of finding the nitrogen at the apex is nearly 100 percent. Left to its own devices, the wave function would shift as time went by, reflecting the increasing probability that the nitrogen atom would be found below the hydrogens. But before the wave function shifts, we make an observation. The act of observation causes the wave function (which, again, describes the probability of the atom's being in this place or that one) to collapse from several probabilities into a single actuality. This much is standard quantum theory, the well-established collapse of the wave function that follows an observation.*
>
> *But something interesting has happened. "The wave function has ceased oozing toward the bottom," as Sudarshan and his colleague Tony Rothman explained, "It has been 'reset' to the zero position. And so, by repeated observations at short intervals...one can prevent the nitrogen atom from ever leaving the top position." If you rapidly and repeatedly ask a system, Are you in this state or are you not? And make observations designed to ascertain whether or not the nitrogen atom is where it began, the system will*

not evolve in the normal way. It will become, in a sense, frozen. As Stapp puts it, "An answer of 'yes' to the posed question [in this case, is the nitrogen atom on top?] will become fixed and unchanging. The state will be forced to stay longer within the realm that provides a yes answer." (Schwartz, 352)

This experiment served as a template for Henry to explain William James's stream of consciousness. If the independent mind continues to ask a question, continues to focus attention in other words, thoughts that well up in the brain can be held in consciousness as long as the focus continues. This is how we are able to call up thoughts. One layer of consciousness, the aware consciousness, observes the subconscious. The subconscious is our belief system. It is in a state of superposition of all possibilities. The aware conscious is indeterminate and acts as the observer.

ALBERT: It sounds good, but I have a hard time believing in some kind of independent mind. I guess, like Descartes, I am an epiphenomenalist. I think that the mind is an emergent property of the brain. I really don't think that the mind can affect matter. The whole notion that we have free will is an illusion. In fact, I recently read an article describing the experiments of Ben Libet.

If I recall correctly, in the 1980s, Libet carried out experiments with volunteers using a device called an electroencephalogram. I'm told that electrodes are placed on the head, and this machine is able to measure brain activity. He asked his volunteers to make a movement— I think it was as simple as raising a finger— anytime they wished. Through careful measurement he found that brain activity began even before the volunteer was aware of the decision to move his finger. This tells me that free will is an illusion. If you're sitting in your chair and you decide to get up and get something to eat from the refrigerator, it isn't really a result of volition. The subconscious mind, which Freud talked about, has already made the decision to get up and go to the fridge. But your conscious mind is not aware of this. Brain activity begins, and fools you into thinking that it was a conscious decision and that you are in control, when in fact, the decision was determined.

NIELS: If that's the case, you deserve no credit for Relativity.

Albert appreciates Niles's wit and laughs under his breath.

ALBERT: Perhaps not, but most of my great breakthroughs seemed to just come out of nowhere. When I was in the patent office, I was just sitting there, when out of the blue it came to me that a person in free-fall would not experience the effects of gravity. As you know that was what led me to General Relativity and a new theory of gravity.

This phenomenon is not all that unusual. One of the greatest discoveries of the twentieth century, the discovery of the DNA helix, came to Francis Creek in a dream. I do believe, however, that much preliminary thinking must precede these events; one must be ready to receive those inspirations from the great void, but the answers to our questions come from somewhere "out there."

> *The picture of a scientist sitting down with his problems and using conscious induction and deductions is as mythical as the unicorn. The greatest insights of mankind have come more mysteriously. The literature is full of insights which have simply come from nowhere. To the mathematician Gauss— like a sudden flash of lightening, the riddle happened to be solved. To Poincare, the brilliant mathematician—without anything in my former thought seeming to have paved the way for it, the solution just came to me. To Einstein on his idea of general relativity—Suddenly the happiest thought of my life came to me.* (Jaynes 43)

A well-known physicist in Britain once declared, "We often talk about the three B's, the bus, the bath, and the bed. That is where the great discoveries are made in our science!" (Jaynes, 44)

NIELS: It's hard to deny the welling up of deterministic thoughts. As you said, one must prepare, but it is your aware conscious mind that observes all of the superpositions that you have been working on. The "aha" experience comes when an observation by your aware consciousness collapses all the possibilities into one particular reality.

Valerie Hunt, a physical therapist and professor of kinesiology at

U.C.L.A., did an experiment with a machine called an electromyograph that is relevant to what you are talking about. The device she used is similar to the electroencephalogram, but the electromyograph measures electrical activity in the muscles. She found that the frequency of the electrical activity of the muscles is around 250 cycles per second, and I believe that the frequency of brain wave activity is around 100 cycles per second give or take. But she discovered another electrical field around the body, which she called the human energy field, and she discovered that it has a frequency of around 20,000 cycles per second. By simultaneously monitoring the human energy field and brain activity, she found that the human energy field responded to stimuli even before the brain registered a response.

ALBERT: I haven't heard about this field. What was her take on it?

NIELS: Hunt would agree with you that the brain is deterministic in nature; that the brain is just a sophisticated computer; however, she believes that consciousness is universal; consciousness exists as a field that the deterministic brain can access.

ALBERT: So, we're back to some kind of dualism, eh. We have the brain over here, and then some sort of entity she calls "mind" in some other place "out there." You are once again reverting to dualism.

NIELS: Not necessarily; if matter can be shown to be similar to mind, then we're right back to a psychical monistic explanation. As you know, matter at the tiniest level of reality has none of the attributes we generally associate with hard cold stuff. At the atomic scale matter has no texture, color, or discrete location in space. Matter tends to look more like mind at this scale. I think it was Sir James Jean who stated that, the stream of knowledge is heading toward a non-mechanical reality; the universe begins to look more like a great thought than like a machine.

ALBERT: I was asking you before about experimental evidence that mind acts on the brain, but we got sidetracked. What have you found?

NIELS: Libet himself had a different interpretation of his experiment than the one you mentioned. He believes that free will is still allowed with the results he obtained. The mind he says has plenty of time to veto these thoughts. Let's take your example of the deterministic thought that says "go to the refrigerator and get something to eat." Rather than acting on the whim, you might decide that it is too early to eat, or that you might want to read a bit longer, or that you are too fat; therefore, you decide to wait. So free will is about vetoing or not acting on those deterministic thoughts. Libet believed that free will is saved in this regard. One part of the brain, the early reptilian brain, is deterministic and the modern brain, namely the frontal cortex, is indeterminate and has veto power over the deterministic aspect of the brain.

So, as I stated earlier, there are really two interpretations to Libet's experiment. One is the Valerie Hunt idea in which consciousness is universal, and the other is Libet's idea that the mind is an epiphenomenon of the brain. In either case, it seems we have the ability to act or not act, and so, we still maintain an aspect of free will.

But back to your question about experimental evidence for volition, I ran into Libet's experiment when I was reading a book by Jeffrey M. Schwartz and Sharon Begley called, *The Mind and The Brain*. Schwartz was a neuropsychiatrist, who was aware of experiments with animals and humans conducted in the 1980s and 1990s that showed the plasticity of the brain. He found that the brain could rewire itself, forming new synaptic connections, as well as generate new neurons.

It's hard to tell what was the biggest surprise to neuroscience: the idea that the adult brain could make new synaptic connections or that entirely new neurons could form in the brain. Both of these notions were totally unexpected. And it's not difficult to understand why.

It was known for a relatively long time that the brain of a newborn undergoes massive rewiring and neurogenesis. At birth a newborn child has around 100 billion neurons with connections to about 2,500 other neurons. Over the next two to three years a flurry of dendritic branching takes place until, on average, almost 15,000 connections are made by each neuron. At three years of age, a child has reached the pinnacle of connections and from that time until early adolescence a pruning process takes place in which some 20 billion connections are lost each day. After age 20, a second pruning process takes place (111).

Because of this diminution, and the known fact that neurons do not divide and replicate, it's no wonder that scientists thought that the adult brain was static and immutable. But there was another more fundamental reason for this stance among scientists that goes straight to the philosophy of determinism which you will appreciate, Al. If the brain is like a computer, that is, if the brain's wiring is determined by our genes, then it would seem unlikely that much change could happen after birth. Any change in the grand design should produce havoc. It would be similar to changing a few lines of computer code and not expecting it to affect the entire program. If the brain is a deterministic system in which connections are established by a grand genetic design, then changing the architecture of an estimated 100 billion neurons with 100 trillion connections should have catastrophic results. But this isn't what was found. It turned out that the brain is a dynamic system, which interacts with the environment and changes throughout life. Not only are new connections made by external interaction, but it was found that new neurons could form in the hippocampus area of the brain from stem cells. Al, I printed out a few pages of some experiments run by the behaviorists as noted in: *The Mind and the Brain,* to get your reaction.

Niels projects the article with his holonet.

> *Experiments like these rekindled Schwartz's interest in the Buddhist tradition. Though a bit of a skeptic, he enrolled in a couple of self-taught courses during his early days of medical school. He was particularly interested in the practical applications of mindfulness training, or what Buddhists call, bare attention and its relation to the volition and the brain. In particular he wondered if mindfulness could help patients diagnosed with obsessive-compulsive disorder. In the late 1980s Schwartz attended a lecture that would change the course of his life.* (1)
>
> *The speaker was widely known in medical circles for her application of the rigorous behaviorist principles to psychological illnesses. "Rigorous," actually, hardly did the behaviorist approach justice. The very first*

paragraph of the very first paper that formally announced the behaviorist creed—John B. Watson's 1913 classic, "Psychology as the Behaviorist Views It"—managed, in a single throw-down-the-gauntlet statement, to deny man's humanity, to dismiss the significance of a mind capable of reflection, and to deny implicitly the existence of free will: "The behaviorist," declared Watson, "recognizes no dividing line between man and brute" (2).

Rarely in the seventy-five years since Watson has a secular discipline adhered so faithfully to a core principle of its founder. Behaviorists, ignoring the gains of the cognitive revolution that had been building momentum and winning converts throughout the 1980s continued to believe that there is no need for a therapist to acknowledge a patient's inner experiences while attempting to treat, say, a psychological illness such as phobia; rather, this school holds that all desired changes in behavior can be accomplished by systematically controlling relevant aspects of a patient's environment, much as one would train a pigeon to peck particular keys on a keyboard by offering it rewards to reinforce correct behavior and punishments to reverse incorrect behavior. The grand rounds speaker, faithfully following the principles of behaviorist theory, had championed a particular method to treat obsessive-compulsive disorder known as "exposure and response prevention."

Exposure and response prevention, or ERP therapy, was a perfect expression of behaviorist tenets. In ERP therapy sessions as routinely practiced, the OCD patient is almost completely passive. The therapist presents the patient with "triggers" of varying intensity (Schwartz, 2).

The theoretical basis of the approach, to the extent that there is one, involves the rather vague notion that the intense discomfort will somehow cause the symptoms to "habituate," much as the intense feeling of cold one feels after jumping into the ocean fades in a few minutes...In the United States, therapists in the forefront of developing these techniques have had patients rub public toilet seats with their hands and then spread—well, then spread whatever they touched all over their hair, face, and clothes. They have had patients rub urine over themselves. They have had patients ring in a piece of toilet paper soiled with a minuscule amount of their fecal material and rub it on their face and through their hair during a therapy session—and then, at home, contaminate objects around the house with it. In other cases, patients are prevented from washing their hands for days at a time, even after using the bathroom (3).

ALBERT: Harsh!

NIELS: Ya, Schwartz was appalled. He recruited OCD patients and began applying his new found knowledge to the treatment of these patients.

In a four-step process of *relabeling, reattributing, refocusing, and re-valuing,* he was able to instill in his patients a method of self-induced therapy (79). Over time these procedures altered brain chemistry, brain function, and produced long term measurable changes in the physical structure of the brain.

In short, he was able to show conclusively that the mind can affect matter. Through pure thought and focus, what the Buddhist practitioners call mindfulness, the mind can affect the brain. His friend, Henry Stapp was able to provide a quantum mechanical description of how this is accomplished, namely the quantum Zeno effect. Through constant observation by the independent mind, the healthful thoughts can be held in consciousness and the unhelpful thoughts can be ignored. With repetition, new neural networks are formed and the brain undergoes a physical transformation.

This can be seen in our everyday experience. If you are paging through a book looking for a specific word, you will be focused on that task and other words are ignored. Or if you drop your keys on the ground you focus on that particular task and all other stimuli are ignored. I had a friend who took up paleontology as a hobby. He told me that he had never before found a fossil until he began to study the subject. Now, because he has attuned his mind to the shapes of fossils, he finds them all of the time. We're back to the believing-is-seeing idea.

ALBERT: On the contrary, it seems to me that we're back to dualism. You say that mind and matter are one, but in the same breath you say that mind affects matter—that it affects the physical brain. I think that I have a much more consistent idea of what is going on here. I think that the idea of free will is illusory, and I think that information and meaning are the basic reality.

It makes more sense than what you are saying. If consciousness is universal as you described it earlier, then why do we need all of those pesky neurons in our brain to produce consciousness? How is consciousness out there different from consciousness produced from within? I thought that quantum theory had gotten rid of the distinction of the observer and the observed.

After the war, I talked extensively with David Bohm when he came to visit me at Princeton. I'm sure you have come across his idea of the implicate order.

NIELS: Yes, I'm well aware of the implicate order and, I must say, I think he was on to something there. Some of his ideas were very obscure, however, and I'm not sure I really understood what he was talking about most of the time. Beside, Al, I thought that you were a determinist. Surely you don't subscribe to such a subjective reality as *meaning* do you?

ALBERT: I'm going to show you that meaning is not subjective as most would think. I'm going to show you, and prove to you, that meaning is objective.

NIELS: This I have to see. Meaning by its very definition is subjective.

ALBERT: You'll see.

Albert pulls out his pipe and prepares to fill it with tobacco.

Niels: Hey Al, you can't smoke in here. They don't allow that in restaurants anymore.

ALBERT: Gees, how is a guy supposed to think?

Albert looking a little embarrassed and reluctantly puts his pipe back into his pocket.

David Bohm (1917–1992) is most noted for his idea of the undivided whole and the implicate order. His alternative theory of the Copenhagen Interpretation of reality called the "ontological interpretation of quantum mechanics" maintains that there can be no distinction between the observer and the observed, the particle and the measuring instrument, nor can there be a distinction between objective science and subjective human experience.

His philosophy, a mixture of modern physics, Buddhist philosophy, and cognitive science, envisioned our everyday reality, the "explicate" order, to be a holographic projection of the "implicate" order, a deeper more fundamental reality.

Bohm's ontological interpretation of reality, which was influenced by Albert Einstein, was an attempt to return to a classical description of reality in which the quantum enigma of the particle/wave duality could be explained in terms of a "pilot-wave" or field.

Bohm and Einstein, neither of whom thought quantum theory to be complete, met after World War II at Princeton University where the two engaged in intense discussion for six months on the topic of quantum theory.

Most notably for our interest here, Bohm reformulated the famous EPR experiment into a conceptually simpler experiment involving the polarization of light.

In the 1950's he fled the United States to England where he found true academic freedom, free from the intrusive and repressive rise of McCarthyism.

Bohm remains one of the most controversial and enigmatic physicists of the twentieth century. Of the controversial Bohm, one physicist said of his ideas, "Some are brilliant, many are obscure, and some are just plain nonsense."

NIELS: I believe that we must admit that reality is subjective.

ALBERT: I'm not willing to concede that point. Last night I got out of bed to go to the bathroom and ran into a chair which left me sprawled out on the floor. I didn't observe the chair, but it had reality nonetheless. That chair was as objective as it could be. I've got a big bruise to prove it. I'm more likely to buy into the proposition that subatomic events are observer dependent than I am to believe that the corporeal world is observer dependent. I still maintain that the moon exists even when I'm not looking.

NIELS: I'm glad that you brought that up, Al. You believe in an objective reality, is that right?

ALBERT: Of course, you know that's the case.

NIELS: And you also believe that mathematics is objective, correct?

ALBERT: Sure, and I'm in good company. Penrose has been making waves along that line.

Roger Penrose, Professor of Mathematics at the University of Oxford, who shared the 1988 Wolf Prize with Stephen Hawking and has penned several popular books including, "The Emperor's New Mind," "The Road to Reality," and "Shadows of the Mind," is well known for his belief that mathematics is objective. He has resurrected Pythagoras's notion that the universe operates according to whole numbers and their ratios. Examples of this are the whole number energy levels of Planck's constant around the nucleus of the atom, and the binary nature of quantum values. Penrose cites the example of Benoit Mandelbrot's set to show that numbers are objective.

> *The Mandelbrot set provides a striking example. Its wonderfully elaborate structure was not the invention of any one person, nor was it the design of a team of mathematicians. Benoit Mandelbrot himself, the Polish-American mathematician (and protagonist of fractal theory) who first studied the set had no real prior conception of the fantastic elaboration inherent in it, although he knew that he was on the track of something very interesting. Indeed, when his first computer pictures began to emerge he was under the impression that the fuzzy structures that he was seeing were the result of a computer malfunction. Only later did he become convinced that they were really there in the set itself. Moreover, the complete details of the complication of the structure of Mandelbrot's set cannot really be fully comprehended by any of us, nor can it be fully revealed by any computer. It would seem that this structure is not just part of our minds, but it has a reality of its own. Whichever mathematician or computer buff chooses to examine the set, approximations to the same fundamental mathematical structure will be found...the Mandelbrot set is not invention of the human mind: it was a discovery. Like Mound Everest, the Mandelbrot set is just there!* (Penrose, 95)

NIELS: Then why is it that you find it so hard to believe in the statistical wave?

ALBERT: Simple, because statistics, unlike numbers, is nothing real.

NIELS: Oh come on!

ALBERT: It's true that one must work with numbers to describe a compilation of events, called statistics, but statistics, like probability, is only our attempt to describe events that have objective causes. Actually, statistics is even less real than probability, because a statistical analysis happens after the fact. For example, in medical studies, we might compile a list of people working on farms, and compare them with people in the general public to find out which group has the most incidences of cancer. We might conclude that people who live on farms are statistically more likely to develop cancer than those in the general public, but it tells us nothing as to why this might be the case. Statistics only attaches numbers to the resulting events. This is the case with nuclear-decay as well. Only after the fact of observing many examples of an element in decay do we assign its statistical half-life, but the fact that it has a probabilistic nature, as I have previously shown, does not provide an explanation for the behavior.

Albert looks at his watch and notices that they have been sitting and talking for nearly two hours. He apologizes to Niels for overstaying his visit.

ALBERT: I saw the weather forecast this morning and it looks like we're in for some nice weather for the next couple of weeks. I'm thinking about heading out to my cabin, and I was wondering if you'd like to come and stay for a couple of days?

NIELS: Your cabin! Where do you have a cabin?

ALBERT: It's over on the Intra Coastal Waterway. It's called Little Egg Harbor. I'm located at the Baker's Acres Campground. You take highway 72 and cross the waterway just south of Surf City, and then head south. I'll give you a map so you can find it. The university gave you a car didn't they?

NIELS: Ya, but how are you going to find the time to vacation?

ALBERT: I'll take my laptop and all my reading material with me. I have to get out of the city every now and then. I get more done when I'm at the cabin. Can you make it?

NIELS: That sounds really fun. I think I can arrange to come for a couple of days. I really would like to walk on the beach and listen to the sound of sea gulls.

Albert: Great. I'm leaving a week from tomorrow, and I think I'll stay for about ten days. I want to talk more with you about some of David's ideas.

Niels: I hope I can swing it. I'll see what I can do and let you know next week. Besides, I can't wait to hear how you believe that meaning is objective.

Albert gets up and goes to the door all the while talking excitedly about the cabin. He tells Niels that he will drop off directions when he sees him tomorrow.

The following week Albert prepares for his journey and heads off to Little Egg Harbor and Baker's Acres. Upon arrival, at his cabin he unpacks and prepares his quarters for the work and relaxation he has planned. His days are spent working, contemplating quantum theory, and walking along the beach. He welcomes Niels with great excitement on his arrival in the late afternoon of a splendid day in early May.

The two rent kayaks and spend the following morning paddling around the harbor, and enjoying the warm sunshine. That evening, after preparing some burgers on the outdoor grill, they relax in some lawn chairs as Albert begins to talk of the time he spent with his friend David Bohm after the war.

Albert: Niels, remember when you spoke of the experimental data in which intention acts on the brain? You talked about the observing mind holding in suspension thoughts welling up in the brain. You more or less placed a divide between mind and matter, even though you claimed that no such division exists. If there is no division, then how is it possible for one to act upon the other? I don't like this idea of two brains: a reptilian brain that is deterministic and the modern brain that is somehow more quantum in nature and is able to veto deterministic thoughts. You have stated that there is only one "world"—the "quantum world". Certainly you don't think that the brain has such an arbitrary division do you? I doubt that any modern scientist would be willing to say that we have two brains: a quantum brain and a classical brain. I believe that David Bohm had an answer for this supposed division between matter and consciousness. He did not like the term "consciousness" at all because it is such a misused and vague term. He preferred the word "meaning". He said that meaning is self-referential. Meaning refers to itself; it understands itself. Meaning is not limited to the mind or the brain. It is a universal concept. His idea eliminates the need for any separate entity such as "mind" that acts back on a material brain. At the same time it rids us of the dualist demarcation that Descartes spoke about. It is a simple and elegant idea.

NIELS: Yes, but it once again leaves us with no free will. If there is no division how can we veto deterministic thoughts that well up in the brain?

ALBERT: You have not convinced me with Libet's experiment. Free will is illusory. David talked about the Freudian ego, which we often equate with the "I" as opposed to the "me." We tend to separate the two. For example, one might say something like, I'm going to drive myself downtown, or I'm going to make myself stop drinking, or I'm going to stop myself from smoking. But the "I" and the "me" are the same person. It's like saying "me" is going to stop "me" from drinking. The idea of an "I" is really an illusion created by the brain. The brain or the ego creates the observer. It is the same illusion you spoke about when you said that the mind observes the brain. They are one of the same.

That is the reason that it is so hard to stop drinking, or smoking, or having sex. You might conceivably veto the idea of getting up and going to the refrigerator to get something to eat, but sooner or later you're going to give in to the "me." That's because there isn't anything other than "me." But you are probably saying to yourself that people do give up drinking and smoking, and of course you are correct. The reason for this is that there isn't just one thought going through the brain at any instant. The thoughts that well up and cause awareness are the thoughts with the highest potential. They have the largest amplitude. So these are the ones we usually act upon. But there are thousands, if not millions of other wave functions going on simultaneously. There are thoughts that say "don't go the refrigerator, or don't light up that cigarette. These cause cognitive dissonance or wave interference. As more reasons build to quit smoking, such as family, or peer pressures, or health concerns the smaller potential thoughts in the brain become larger. At some point the "don't smoke" wave function becomes larger than the "do smoke" wave function and we are finally able to quit.

As Albert talks he loads his pipe, strikes a match, and takes a deep draw.

NIELS: You obviously haven't reached that point yet.

Albert ignores Niels' slight.

ALBERT: Now it might well be that the frontal lobes, which give us the ability to do complex functions and give us the sense of "I," are a recent evolutionary advance, but all parts of the brain produce deterministic waves: the reptilian brain and the more recent complex frontal lobes.

Other animals besides humans experience the sense of "I." Researchers have tested this by placing mirrors in front of animals to find out how they react. Some perceived their own reflections as other animals as if staring through a window. Others, such as dolphins and some apes, realized that they were in fact looking at a reflection of themselves and would posture in front of the mirror as if they were examining themselves. No one can doubt that most, if not all, animals are conscious, and they experience an inner world as we do. One needs only to watch the lion in pursuit of its prey to realize that their interpretation of the sensory signals of sight, smell, sound, and touch must be very similar to our own.

Yet, there certainly is a difference between primary consciousness and that of higher awareness that we possess. A statement that illustrates this point is that you might wonder if your dog is conscious, but it is very unlikely that your dog wonders if you are conscious. We and a few other animals seem to be uniquely equipped to ponder. The deterministic brain is the Occam's razor of consciousness for two reasons. First, it relieves us of the task of trying to explain some nebulous concept as the mind. Secondly, it brings back the idea of cause and effect which quantum theory has attempted to eliminate completely by replacing it with randomness.

NIELS: How so?

ALBERT: David thought that meaning is the ultimate source of cause and effect. Meaning is enfolded in matter and energy. If meaning is enfolded within all matter and energy, in what he called the implicate order, then there is no separation between mind and matter, nor can objectivity and subjectivity be discrete. In fact, if the universe is organized such that meaning is at the very core of reality, than we must come to embrace a very counterintuitive concept—the universe is contextual, yet objective. An interesting consequence of this idea is that it brings back cause and effect into quantum theory.

NIELS: So you are saying that if the entire universe is contextual, if subjectivity is ubiquitous in nature, then by definition subjectivity becomes objectivity!

ALBERT: Yes, that's right.

NIELS: This is quite extraordinary. From the ancient Greek philosophers, to the idealist philosophers, to the quantum revolution, much ink has been spilled trying to determine how much of reality is subjective and how much is objective. You have now rendered the distinction null with one fell swoop. Be specific, how does that bring cause and effect back into quantum mechanics?

ALBERT: Remember that David Bohm equated consciousness and meaning. Therefore, if you claim that it takes consciousness, or as David would say meaning, to collapse the wave function, then consciousness could have an objective physical effect.

NIELS: If consciousness or meaning as you prefer to call it is objective, does that necessarily mean that it is deterministic? And if it is deterministic, are you claiming that we can have no free will?

ALBERT: You could say that we have an illusion of free will simply because our brains are the most highly organized matter in terms of meaning. The speed of operation is the only demarcation between the perceived free will we have, and the deterministic attributes of nature. Nature slowly and relentlessly moves toward meaning, from the alpha to the omega; whereas, our brains are able to find meaning extremely fast, giving us the illusion that we uniquely possess free will and giving us the illusion that nature is random and meaningless. Yes, I believe we must conclude then that nature, and therefore, meaning, is objective and deterministic as well.

NIELS: The chair that left you sprawled on the floor was objective and you are now saying meaning is objective. I think I'm starting to understand your line of reasoning.

Oh yes! I remember now. This is the carrot you dangled in front of me to get me here this weekend. You said that you would prove beyond any doubt that meaning is objective. I'm all ears. Go ahead, try to convince me, I'm listening.

ALBERT: Ok, that's a challenge. I shall not only show you that meaning is objective, but I will show you that quantum probability is objective. Would you hand me your cup, please?

NIELS: Alright.

Niels slides his cup over to Albert with a look of skepticism.

ALBERT: Ok, now I'll take your cup and my cup and turn them upside down. Now, Niels, if you wouldn't mind, would you close your eyes for a moment?

NIELS: Sure.

ALBERT: Ok, you can open your eyes and look. I placed a pea underneath one of these cups. Can you tell me which cup the pea is under? Before you take a guess, we know that you have a 50/50 chance of guessing correctly. I know which cup the pea is under, but you don't, so this is a case of classical subjective chance. The pea is definitely under one cup or the other, but you don't know which, so the situation is one of subjective meaning.

Now I'll ask you to guess.

NIELS: I'll guess this cup over here has the pea.

ALBERT: Well, you guessed correctly, and now your sense of meaning has changed, but nothing has changed about nature. The pea was here all along. It was simply your lack of knowledge about its location that changed. Since I knew where the pea was and you didn't, your choice was a subjective choice.

But now, for the sake of a thought experiment, let's suppose that our pea becomes a single atom. Let's set up a typical thought experiment using our very small quantum pea. Suppose I have a source that ejects these imaginary miniature peas, and I direct these peas at a half-silvered mirror, just like in the typical photon experiments. Now the pea has a 50/50 chance of going through the half-silvered mirror and going into cup "A", and a 50/50 chance of it being deflected and ending up in cup "'B". Do you have any questions up until now?

NIELS: No, so far this is the typical photon experiment. Go ahead, Al. I'm following along. (*Figure* 6)

ALBERT: Ok, Good. You and your Copenhagenists friends maintain that the "pea" goes into a superposition of states and that the pea is both nowhere and everywhere simultaneously, is that correct?

NIELS: Yes.

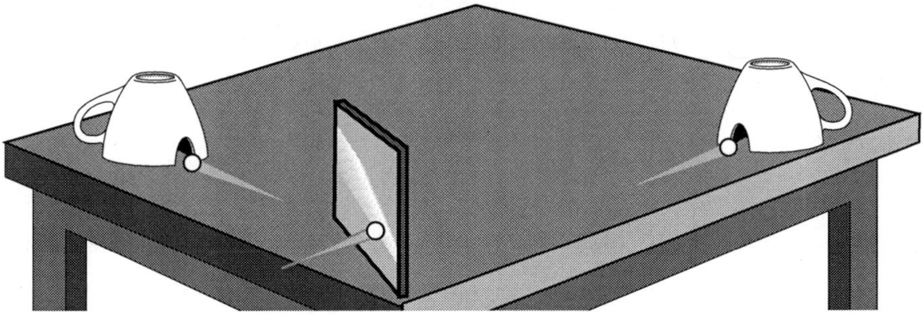

Figure 6.

ALBERT: The pea is both in cup "A" and cup "B" and **not** in cup "A" and cup "B". The lack of knowledge isn't limited to you or to me, it is limited to everyone, and to nature herself; therefore, in this case quantum chance is objective. The instant I look under one cup or the other the wave function, as you call it, collapses and the pea materializes under cup "A" or cup "B." The observation changed reality itself. The observation changed the meaning of the situation for one and all; therefore, chance and meaning are inherently objective aspects of reality.

Niels, you keep reminding me that the electron or the photon statistical wave isn't anything "real." I believe that we have to re-evaluate what is "real." Meaning is real; meaning is objective. And I believe, like Pythagoras that numbers are objective and real; meaning is objective and real; the probability wave is objective and real.

NIELS: I see your reasoning Al. You have admitted that the observation changed the experiment. We still have the observer affect to deal with. But yes, I'm impressed.

ALBERT: Are you glad you came now?

NIELS: Sure. Paddling around the beautiful harbor would have been reward enough, and now this. But I have one concern. Previously, you suggested that probability and statistics are nothing real, and now you are telling me that the probability wave is real. Isn't this a contradiction?

ALBERT: No, not the way I'm thinking of this. One can never use probability or statistic to explain why something happens. Probability can only describe what one observes, but it is not a basic law. It isn't probability in and of itself that is real. Probability only describes the likelihood of something happening based on past observations. Probability, in and of itself, is nothing objective, but the probability wave is objective. To alleviate any confusion we should re-label the probability wave and simply call it the particle wave. The particle wave is objective and it is real, and it is deterministic.

NIELS: You slipped in that word "deterministic." I'll have to think on this for a bit.

Al, is the objectivity of mathematics more basic than the laws of nature, such as gravity and the electromagnetic force? Is mathematics even more basic than the laws of thermodynamics?

ALBERT: I think that the forces *are* mathematics and nothing more. We pay homage to the great Isaac Newton for his three laws of motion and his law of universal gravitation. He discovered that every object in the universe attracts every other object, and he discovered a very interesting mathematical law that defined this attractive force: the force between these objects was directly proportional to their masses and inversely proportional to the square of the distance between the objects. This famous formula is credited with providing a description of all planetary motions and it has given scientists and engineers the mathematical know-how to calculate the trajectory of objects and the velocities needed to launch men and machine into orbit. It is no wonder that Newton is often thought of as the greatest scientist of all time.

NIELS: You're being modest, Albert.

ALBERT: As you know, the inverse square law can be applied to the electromagnetic charge as well. Just as with the gravitational force, the electromagnetic force falls off with the square of the distance. Can this be merely coincidence? Nature seems to operate according to mathematical laws, and it is these laws that are meaningful. Kepler, Newton, and I produced three different theories that describe virtually the same thing. My description of gravity states that gravity warps space, but it doesn't tell us anything about what gravity is. We are no closer to understanding what gravity is than the ancient Greeks, so is there any sense in going beyond a mathematical description of gravity?

I remember when Planck and I would argue about whether his ground-breaking formula $E = h\nu$ was anything more than a mathematical description of what happens when a black body radiates energy. As you know he found that he could explain how a body radiates energy in discrete packets or units of his mathematical constant. And I found that light itself was made of quanta, yet he was never convinced that this was really the reality of the situation. He always thought that it was only a mathematical trick to explain what happens. And though I played antagonist at the time, I was not really sure that I was right. But more to the point, I sometimes wonder what came first, the mathematical laws, or the behavior of the atoms? And if it was the mathematical laws that came first, were they already in place or did we create them? In other words, did nature conform to the mathematical description we produced? Did we beat nature into line as physicist and historian Thomas Kuhn claims?

> *In the eighteenth century, John Dalton's new chemical theory of compounds featured a completely different set of assumptions from those which guided chemists of his day. When Dalton searched the chemical literature to find evidence for his theory, he found some examples of reactions that fitted his hypothesis but many that didn't. For instance, one experimentalist famous or the accuracy of his measurements had analyzed the oxygen/copper proportions in copper oxide as 1.47:1. Dalton's atomic-weight theory demanded a ratio of 2:1. Kuhn says of this: (Briggs, 30)*
>
> *"It is hard to make nature fit a paradigm. That is why the puzzles of normal science are so challenging and also why measurements undertaken without a paradigm so seldom lead to any conclusions at all. Chemists could not, therefore, simply accept Dalton's theory on the evidence, for much of that was still negative. Instead, even after accepting the theory, they had to beat nature into line, a process which, in this event, took almost another generation. When it was done, even the percentage composition of well-known compounds was different. The data themselves had changed."* (Briggs, 31)

ALBERT: The numerological interpretation of reality implies that the universe can be understood using the most meaningful and powerful tool of human reason—mathematics. Pythagoras and the mystical organization he founded, the Pythagorean brotherhood, felt that numbers belonged to the highest realm of existence. Since they believed that reason alone was sufficient for understanding the universe and mathematics was a tool to further this reasoning process, one would tend to believe that mathematics would be subjective, but Plato said that the mind was also a mirror of the universal realm. So in that respect the mind itself is objective and, therefore, numbers are objective as well. If the mind is a miniature universe, as Plato thought, and meaning itself is objective, as we have shown, then the argument as to whether mathematics is objective or subjective dissolves into obscurity. Numbers are a creation of the mind which is meaningful, but as we saw, meaning is a part of objective reality; therefore, numbers must be objective.

It seems quite possible that when meaning is expressed in numbers and mathematics, nature responds by aligning herself with this mathematical reality. Buddhists are fond of saying that *consciousness* is the ground of all being. David Bohm and I would modify that a bit to say that *meaning* is the ground of all being. So it is quite possible that when we make mathematical theory that is consistent, and meaningful, it becomes an objective part of the universe as a whole. As you can see, the dividing line between objectivity and subjectivity gets blurred.

NIELS: You have removed an obstacle that has divided us since the 1920s. By demonstrating that there can be no divisions between objective and subjective reality, you have in turn united the dualism of the observer and the observed. And you have laid out your argument in a much clearer way than I have ever been able to do. I commend you, Al. I think that this surpasses your space/time synthesis.

ALBERT: Yes, and when I saw this unity of nature it gave me the same sense of euphoria I felt when I formulated special relativity.

NIELS: The ideas you have presented reminds me a bit of the *axiom of internal relations.*

ALBERT: Refresh my memory.

NIELS: During the nineteenth century the philosophy of monism flourished. It was believed that attributes which we ascribe to an entity are not inherent in the entity itself, but internal to the observer.

The Axiom of Internal Relations

> *If we consider the statement: "This emu's egg is bigger than that hen's egg," it would normally be said that the fact which the sentence expresses consists of two separate entities—the emu's egg and the hen's—and of a relation, namely, the relation of being greater than, which is asserted to hold between them. The two eggs, we should say, are distinct from each other, and distinct also from the relation which relates them...Taking the instance given, the advocates of internal relations would point out that, unless a hen's egg were smaller than an emu's, unless it were browner than a pigeon's and larger than a sparrow's, unless it were more brittle than rubber and more oval than a billiard ball, it would not be the egg it is. Hence, its attributes of "being smaller than," "being browner than," "being larger than," "being more brittle and more oval than" literally constitute its nature. If it did not possess these attributes, it would be a different egg. But each of the attributes in question presupposes the existence of something else to which, in virtue of its possession of the attribute, the egg is related. If there were no emus' eggs, pigeons' eggs and sparrows' eggs, no rubber and no billiard balls, these attributes of a hen's egg could not be its attributes. Therefore the relation of the hen's egg to other things and, by consequence, the existence of these other things, constitutes part of the being of the hen's egg...Change a thing and you change its relations, you change, therefore, everything in the universe.* (Joad, 413, 414)

NIELS: This axiom is much broader in scope than just saying that our reality is a matter of perception. This axiom suggests that meaning is inherent in nature. Change anything in the universe and everything else changes in relation. This is the objective understanding of meaning that we've been discussing.

ALBERT: Yes, and we could say that meaning is relative just as space and time are relative.

NIELS: But you said meaning is objective. Relativity is a subjective proposition.

ALBERT: That's true. Remember that space and time are both relative concepts, but the more universal concept which I synthesized into space/time is an objective concept. The Emu's egg is a relative concept, but the more universal concept, the axiom of internal relations, which is a statement about the universe in general, is objective. It is a property of the universe at large.

The sun has set and the light begins to fade. Albert lets out a long sigh...

ALBERT: I really miss seeing the stars.

For the past decade NASA in collaboration with the United Nations has been staging regular launches of an orbital vehicle to release tiny reflective particles over the tropical zone in high earth orbit. The purpose of the particles is to reflect into space some of the sunlight reaching the earth in order to reduce global warming. The particles have been successful in reducing the amount of sunlight reaching Earth, but they have caused other problems. During the night when the sun is on the opposite side of earth, the particles catch the sunlight and reflect it back to earth causing a night time glow that diminishes the possibility of seeing stars.

ALBERT: If man had never had the stars to gaze up at, civilization would have never advanced. How would we have even guessed that we were part of something larger? I remember when I was in Bern, Switzerland; I had just gotten my job at the patent office. Maurice Solovine, Conrad Habicht, and I formed a little club that we called the Olympia Academy. We would often hike up into the Alps on summer nights, find a nice spot to throw out a blanket, watch the stars, and talk about girls and physics. We would talk and laugh until the sun came up. Those days were the happiest of my life. Wisdom gained is never worth the loss of youth.

NIELS: Last week you talked about how you fled your unhappy life in Germany and went to Italy to join your father and uncle in business. What was life like in Italy?

ALBERT: I liked everything about Italy. I liked the casual life style and the friendly people.

NIELS: Did you study for college like you promised your folks?

ALBERT: Yes, I was able to completely devote myself to my studies. The love of learning had returned, and I would bury myself in physics text books with total concentration for hours at a time. That fall, I felt I was ready to enter college even though I was two years younger than the requirement. A friend wrote a letter to the Polytechnic recommending me to take the entrance exam. I went to Zurich to plead my case with the administrator and even though he was very skeptical he acquiesced. In October of 1895, I returned to Zurich and took the exam.

NIELS: Did you pass?

ALBERT: No. My scores in mathematics and the physical sciences were adequate, but I fell short of the mark in subjects like literature, zoology, and politics. I was never one to give up, so I decided to go back to school in Aarau, Switzerland. That turned out to be one of the best decisions of my life both academically and emotionally. When I moved to Aarau, I stayed with Jost and Rosa Winteler and their seven children. Jost taught history and Greek at the school I attended. They were social democrats, and their liberal progressive and social views suited me perfectly. And best of all, they were pacifists. It was a warm and nurturing environment and that is exactly what I needed at that time in my life. It wasn't long after moving in with them that I fell in love with their daughter Marie. She was two years older than I was, but both her parents and my parents approved of the relationship. They thought we were a cute couple apparently. Not only was their daughter Marie my first girlfriend, but Anna, another daughter, ended up marrying my best friend Michele Besso, and their son Paul married my sister Maja.

The school I attended gave me the freedom to explore the far reaches of my imagination and it rekindled my thirst for learning. We were treated as individuals and were encouraged to approach our work using conceptual thinking and visual imagery. This was a far cry from my experience in Germany. I once again began to ponder such things as what it would be like to ride alongside a beam of light.

NIELS: Did you eventually get into the Polytechnic?

ALBERT: Yes, I took the entrance exam again in the fall of 1896, and I passed, but my college career was abysmal.

NIELS: What happened?

ALBERT: My old bad habits returned. I had a couple of professors I detested, namely Heinrich Weber and Jean Pernet who held equal derision for me. The situation at the institute began to feel like my schooling in Germany, so I started to rebel. I cut classes and I completely sloughed off math. I made a terrible misjudgment in those years. I thought that advanced mathematics was unnecessary for understanding physics; a miscalculation that haunted me for the rest of my life. Weber and I got along well at first, but he was part of the old guard and was too entrenched in the physics of the past. My physics courses seemed more like physics history classes. Weber felt that I was disrespectful of the establishment and we soon became mortal enemies. And my relationship with Pernet was no better. He gave me the lowest possible grade in experimental physics.

NIELS: I'm surprised you did poorly in experimental physics.

ALBERT: Well, I think it was partly due to the explosion I caused in the laboratory that nearly blew off my hand. Pernet gave me instructions on how to perform the experiment, but I felt that I could do it a better way. I wadded up the instructions he gave me and threw them in the trash.

NIELS: That sounds like you.

ALBERT: Things were not going well at the institute, and to make things even more complicated I became infatuated with Mileva Maric who was the only female in our class.

NIELS: Your first wife?

ALBERT: Yes. I couldn't concentrate on anything but Mileva and as a result my studies suffered. I would sit in class and imagine what it would be like to be with her. After classes Michele Besso, Mark Grossmann, Mileva, and several other friends, would meet at the Café Metropole on the banks of the Limmat River and discuss philosophy and physics. Mileva became one of the guys. I guess she had no other choice in that male dominated world. She was a really good sport and she could hold her own. Besides that, she helped me with my math. We would all talk and laugh into the late evening, and then Mileva and I would go to my apartment and we'd roll around in bed.

NIELS: You were a bit wilder than I was. What did your family think of Mileva?

ALBERT: They didn't approve, especially my mom. She wondered why I would want to be with an ugly Serbian peasant girl who was sickly and walked with a limp, when I could have anyone I wanted. She fell into depression and nearly disowned me over the situation.

NIELS: Why did you find her so attractive? Was it because she was the only girl in class?

ALBERT: No, I had plenty of other opportunities outside of class, and I even visited my old girlfriend Marie Winteler a time or two, but to have a girlfriend who I could talk to about physics was very alluring.

NIELS: How were you able to graduate if you did that poorly?

ALBERT: As time went on, I began to buckle down, and thanks to Marcel Grossmann, who lent me his meticulous math notes, I was able to cram for my exams and eeck out a respectable grade.

The next hurdle came when I had to do a research paper that would pass Professor Weber's approval. I was at his mercy, and so, he decided to take advantage of the power he had over me. I first proposed an experiment that would measure how fast the earth was moving through the ether, but he said that experiment had already been done by Michelson and Morley. So I came up with another idea and he rejected that idea as well. It became apparent that he was not going to accept any proposal that did not have to do with the theory of heat conduction which was his specialty. He had me over a barrel, so Mileva and I decided to team up and we proposed doing an experiment that had to do with heat conduction, for which neither of us had any interest. And of course he approved the proposal. But, he wasn't through with me yet. He was enjoying watching me squirm for my grade. After submitting the work, Weber said that the paper wasn't formally correct, so he made us rewrite the entire paper. He gave us a passing grade, but Mileva and I got the two lowest grades in class.

NIELS: Did Mileva get her degree?

ALBERT: No, unfortunately not, and I feel really badly about that. She returned a couple of years later and attempted to graduate, but by then I had gotten her pregnant and school was more than she could handle with the rigors of the pregnancy, so she was forced to drop out.

Several minutes of silence pass as Albert takes a long contemplative draw on his pipe.

Niels, I was always cast as a loner by the press and some biographers, but time and time again it was my friends who pulled me through. Whether it was my family, who encouraged education and music; Max Talmud, who brought me science and philosophy books to study; a note from the family friend that got me out of the school in Germany; the Winteler's, who nurtured my social and humanistic needs; the letter from a family friend that got me into the institute; or the academic help from Grossman, Marie, and others. It was my friends who got me to where I am today. I've never publically acknowledged that. I've been really fortunate to have had such good friends. I had a lot of advantages that others didn't get.

NIELS: What did you do after graduation?

ALBERT: I was the only one from my graduating class who was not recommended for a job. Traditionally, I should have been the prime candidate to become Weber's assistant, but he hired a couple of engineering students instead. And my mathematics professor, who would later help me formulate a mathematical description of special relativity, Herman Minkowski, called me a lazy dog. I had burned all of my bridges at the institute. It was a full nine years after I graduated and four years after publishing my five papers in 1905, that I was able to find an academic job. I did eventually land a job at the Bern patent office thanks to Marcel Grossmann, who informed me that there was to be an opening there. His father was a friend of the director of the patent office and he said the job was mine if I wanted it.

NIELS: How did you survive all of those years without a job?

ALBERT: Looking back to those days, I sometimes wonder how we got by as well. I did some tutoring that brought in enough for us to live on, but the truth is that we didn't need much. We lived the Bohemian life-style. Honestly, I don't know how we got by, but Mileva and I were very happy.

NIELS: Was this the time during which Mileva became pregnant?

ALBERT: Yes, I had moved to Bern in anticipation of getting the job at the patent office and I got Mileva a hotel room in a town close to Bern. But, I just didn't go to see her that often. I dreamed up many excuses such as not having enough money, but really, I just think I didn't like to see her pregnant. To me her pregnancy symbolized the end of youth and independence. I wasn't ready for that kind of responsibility. And to be honest, it was a sexual turn-off. It was selfish of me, I admit, but there were other practical considerations. I couldn't afford to be seen in public with a pregnant Mileva. My mom would have died if she found out. Besides, I might not have gotten the job at the patent office if I had shown up with an illegitimate child. It just wasn't in anyone's best interest to let the secret out. Mileva felt abandoned, and so she went home to her parent's house in Novi Sad to have the child.

NIELS: Was it a boy or a girl?

ALBERT: A girl. She named her Lieserl. Mileva's friend adopted the baby and Mileva moved back to Zurich to stay until I was able to help support her and possibly get married.

NIELS: Did you ever see Lieserl?

ALBERT: No.

NIELS: What was the job at the patent office like?

ALBERT: It is hard to describe how grateful I was to have a real job. It gave me a sense of worth, and of course, I really needed the money. Most people who held a patent clerk job felt it to be extremely tedious, but I looked at it as an opportunity to work on my own ideas. After being at the patent office for a month or so, I became efficient enough to get all of my work done by around noon each day. The rest of the day I worked on my own projects. If one of the supervisors would come by, I would shove my work into my desk and pretend to be working diligently on the matters of the patent office. That job gave me the security and discipline I needed. I had to be there at least eight hours a day whether or not I had enough work, so it allowed me a perfect opportunity to think about physics. Those seven years at the patent office were the most productive of my life.

NIELS: Especially 1905.

ALBERT: Ya, not only did I churn out those five papers, but I was able to complete my doctoral thesis and get it published the following year.

NIELS: Tell me about those five papers. How did those ideas come about?

ALBERT: Well, few people are aware of the fact that by the time I got the job at the patent office I had already published five papers that got published in the German journal *Annalen der Physik.*

NIELS: I didn't know that either. What were those papers about?

ALBERT: Those papers dealt mostly with the reality of atoms and molecules. They didn't really stir up much interest, but they did give me a foothold in the academic world.

NIELS: So what was the impetus behind your 1905 papers when you worked at the patent office?

ALBERT: Well, my first paper was on the photoelectric effect, which eventually won me the Nobel Prize in 1923. As you remember, Max Planck had mathematically explained why a blackbody emitted light in precise quantities or units which later became known as quanta. Before Planck came along, no one could explain why, when monosynchronous light was shone on a body that absorbs all frequencies of light, the heat and light emitted from the blackbody fell mostly in the midranges. This had no classical explanation. It was thought that as the amplitude of the light shone on the blackbody increased, the energy emissions in the form of heat and light should be of a higher frequency. But this isn't what happened. Planck pondered the situation for some time and tried various mathematical formulations until he happened upon a solution that fit the reality of the situation. He noted that if the emissions were mathematically divided into discrete units the emissions could be explained. He then made the intuitive leap that the emissions were not dependent upon the amplitude or brightness of the light, but solely upon the frequency of the light: The higher the frequency of the light, the more energetic the emissions of electrons. He came up with the formula, $E=hv$. Energy is equal to the frequency times a constant which later became known as *Plank's constant.*

When I read his work, it occurred to me that if the light was absorbed in whole units by an atom, then light itself must be composed of quanta or photons as they were later known. This, of course, was a huge leap, because Thomas Young's diffraction experiments with light nearly a century earlier convincingly demonstrated that light was a wave, a departure from Newton's belief that light was a particle. So things went full circle concerning the nature of light. As it turned out, everyone was correct because you and Heisenberg discovered that light was both particle and wave.

NIELS: Yes, but our wave/particle idea was supposed to serve only as a metaphor.

What was your second paper about?

ALBERT: My second paper was my dissertation and it had to do with how one could more easily determine Avogadro's number by measuring the effects of the change in viscosity when a diffusion of sugar is added to a solution. It allowed one to determine the size or gram weight of a sugar molecule for example. That paper was probably the most obscure of the five I published that year.

NIELS: What came next?

ALBERT: Let's see now. If I remember, I had just finished my doctoral thesis, and I was reading of Robert Brown's work with pollen grains. He noticed that when pollen grains in a liquid are observed under a microscope they jiggled. I immediately made the assumption that this could be attributed to the new atomic theory. Well, I should say the newly revised atomic theory. The Greek philosophers Leucippus and Democritus had proposed centuries ago that matter was composed of atoms, but as yet, no one had actually seen atoms. I thought it quite likely that the jiggling might be the result of the random motion of atoms in the water. So, I wrote up a paper and sent it off to the *Annalen der Physik*. After my paper was published most skeptics were silenced, and the theory became generally accepted by scholars.

NIELS: Ya can't have quantum theory without atoms.

ALBERT: It was my fourth and fifth papers that were my most original work. As far as I know, no other scientists were working on anything close to what I was working on with special relativity, and I don't think that most people even realized that there were any paradoxes concerning the speed of light relative to the observer. So, my paper hit the world like a bombshell. I remember that when I wrote my paper *On the electrodynamics of moving bodies,* I felt almost embarrassed that I had so few notations. But there just hadn't been much previous work to cite. The truth of the matter is that I did the work by the platonic method, through pure reason. It simply was a matter of lying on my bed and thinking about my boyhood paradox of flying alongside a beam of light. What would it look like? I came to realize that it didn't make sense for beam of light to be frozen in place. This would be in violation of Maxwell's equations, and if light is the same for all observers in state of constant motion, then in the equation, *Distance* = *Rate* x *time,* something had to give, and that would have to be distance and time. Let me give you an example, Niels. If you are rushing toward a beam of light at around 50% of the speed of light in a spaceship, and you measure the speed of light racing toward you, its speed would be about 186,000 miles per second, and if you turn your spaceship around and raced away from the beam of light and then measured the speed of light overtaking you, you would once again measure its speed to be about 186,000 miles per second. This is of course very counterintuitive, but if we assume that this is correct, as it must be, then space and time must be relative concepts in order to balance the equation. The idea is merely an extension of Galileo's idea of relativity. I took it one step further to include light.

When I am out and about I hear people proclaim that I said "all is relative", but this is just not the case. In fact, I was originally going to call my idea *the theory of invariance,* because I wanted to show that the laws of physics are invariant for persons in the same frame of reference—observers in a constant state of motion.

NIELS: 1905 was a magical year. I don't think that the world will ever see anything like it again—certainly not from one man.

ALBERT: Well, thank you, Niels. Everything just seemed to come together for me.

My fifth and final paper that year was more or less an appendix to the paper on relativity. Several considerations brought me to the idea that matter and energy are equivalent. First, it occurred to me that if atoms absorb and emit energy in the form of light particles, then it should change the mass of the atoms. I made the intuitive leap that matter and energy were two aspects of the same thing. I then took Newton's equation, force equals mass times acceleration squared, and modified it slightly.

Force requires energy, so I substituted E for F in Newton's equation. Acceleration of course is the speed of light, because all radiation in the light spectrum moves at the speed of light. So, Energy is equal to the mass of a particle times the speed of light squared which is an absolutely enormous number.

NIELS: What happened after your 1905 papers were published?

ALBERT: There was a deafening silence from the academic community, and it was Max Planck who eventually brought out the importance of the papers in some of the articles he published. You might say that he rescued me from obscurity. Soon my life became a whirlwind as my fame spread around the world. Unlike most scientists at that time, my fame wasn't just limited to academic circles. Headlines in newspapers around the world reported on the topic of special relativity. It captured the public's imagination. Most didn't understand just what relativity was about, but they knew it was both strange and revolutionary. That, coupled with my wild hair, cast me as the stereotypical mad scientist. But with all the fame I was still not able to get an academic job. I was so desperate that I answered an ad for a mathematics and geometry teacher at a high school in Zurich. There were twenty one applicants for the job, but I wasn't even among the finalists for the position. Finally, in 1909, four years after the publication of my papers, I landed a teaching position at the University of Zurich. This was exactly what Mileva and I had been hoping for. It was a dream-come-true, because both of us loved Zurich and many of our college friends still lived there. If I had never received another offer, we could have lived out our lives there and been perfectly happy. But life happens and soon more lucrative offers started coming in. Just six months after accepting the position at Zurich, I was offered a full professorship at the University of Prague. My wife begged me not to leave, and my students at the University of Zurich signed a petition to get the faculty to keep me around, but the money and the prestige were just too alluring. I took the position in Prague with the stipulation that I could keep my Swiss citizenship.

Niels, I tell everyone that my reasons for leaving Zurich were for money and prestige, but just between you and me, there was another reason. During my time in Zurich, I began corresponding with my second cousin Elsa Einstein, who was living in Berlin. Part of the reason I took the job was to get close to Elsa. I made several excursions to Berlin to see her and, of course, we became romantically involved. I felt very guilty about it, and subsequently tried to break things off. The whole situation in Prague was a disaster in every way for Mileva and me. We hated Prague. The city was filthy, the water was undrinkable, the air was choked with soot from the coal fires, the people we met were snobbish, and on top of all that I had become unfaithful to Mileva. So when the offer came to teach at Zurich Polytechnic, which had just been upgraded to full university status, we jumped at the chance to move back to Zurich.

NIELS: Zurich is a beautiful city.

ALBERT: Yes, life couldn't have been better. I loved teaching at the University, Mileva and I were on good terms again, we were living in a luxurious six-room apartment with a view of the mountains, and thanks to Marcel Grossmann's mathematical help, I was able to make good progress on my attempt to generalize my theory of special relativity.

NIELS: How long were you there this time?

ALBERT: I had everything I wanted and I was very happy, but I began to get restless again because of my feelings for Elsa. It might not have ever come to anything were it not for a fateful event that transpired in July of 1913.

NIELS: What happened?

ALBERT: Max Planck, Walther Nernst, and their wives arrived on the train with a lucrative offer that would include a position at the Prussian Academy of Science, a professorship at the University of Berlin, and if that weren't enough, I would become the director of a new physics institute. All of this was very attractive, but there were a couple of perks that sealed the deal for me. They assured me that I would have no required teaching duties at the university, and I would be allowed to keep my Swiss citizenship. Not having any teaching responsibilities would allow me to pursue my own work, and keeping my Swiss citizenship was a lifeline in case I wanted to bail out and return to Zurich.

I told Max and Walther that I would have to think about it, but deep down, I knew I could not turn my back on such an offer. Besides, I wanted time to persuade Mileva that it would be in our best interest to take the position.

NIELS: Was Mileva persuaded?

ALBERT: No, not really. The truth is that the move was only in my best interest. The main reason I decided to take the job was so that I could see Elsa.

NIELS: Why were they so intent on getting you to come to Berlin?

ALBERT: The Germans were gambling on me as they would on a prize-winning- hen, but I didn't know if I could still lay eggs (Isaacson, 180).

The two begin to clean up and prepare to call it an evening.

NIELS: Well, I should head back tomorrow. What do you say we get up fairly early and do some more kayaking around the bay? I could use the exercise before I take the long drive back.

ALBERT: Are you sure you can't stay another day?

NIELS: No, I'm supposed to do another lecture at the U. If I accepted all the invitations to lecture, I would be doing it every day. It's very odd, Al. They complain that I'm a poor lecturer, that my lectures are boring and inaudible, that I spend too much time lighting my pipe. They say only the students sitting in the front row are able to comprehend my meaning, and yet they can't get enough of my lectures.

ALBERT: Well, you are a celebrity now and this country thrives on that.

The two turn in for the night. The next morning, after having breakfast that Albert prepared, they walk down to the dock and procure a couple of kayaks for a paddle around the bay.

NIELS: Ah, nature at its finest.

ALBERT: Ya, this is my favorite time of day. Here's some bread if you want to feed the sea gulls.

Albert gives Niels some slices of bread and Niels tears off pieces and tosses them into the air as the sea gulls snatch them in mid-flight. After a couple of hours the two return to the dock and check their boats at the rental service. Niels thanks Albert for the great time and reluctantly leaves the beautiful resort for the city. Albert returns a few days later and the two find themselves back in their routine at Princeton University.

Albert has decided to visit Niels in his office next door. Niels had been feeling that his office was too cold and sterile in appearance so with the help of a design specialist he made some changes to give it a more cozy feeling like Albert's office. Niels' office is large and spacious with large wooden framed windows that still have the weighted rope mechanisms for easy opening and closing. One wall of his office has shelves from floor to ceiling filled with books, and adjacent is a drafting table and a computer desk. He has added some throw rugs to cover the tile floor. Pieces of modern and impressionistic art by Magritte and Picasso hang on the wall, and an old rusty looking horseshoe is crudely nailed above the doorway. Niels hears a knock at the door.

NIELS: Come in. Hey, good to see you. I must thank you again for the weekend out at your cabin. I didn't know how much I needed that. I feel completely rejuvenated.

ALBERT: It was my pleasure. I needed it as well. What's this?

Albert is looking about and sees the horseshoe above the door.

ALBERT: Well, look at that. I'm surprised that such a renowned scientist as you is superstitious. Surely you don't believe that a horse shoe brings good luck?

Niels: No, but they say that it works even if you don't believe.

Albert laughs, not knowing for sure if Niels is serious.

ALBERT: That is very surprising in light of your views about the placebo effect. You said that believing doubles the potency of the effect; just think how lucky you would be if you believed!

Albert notices the Picasso hanging on the wall.

ALBERT: I like the Picasso. The cubist style has always intrigued me.

NIELS: It probably reminds you of your block-time theory of the universe.

The two study the painting for a few moments, and then Niels launches into a story:

NIELS: Al, one day Picasso was traveling cross country by train, admiring the scenic Spanish countryside from his window. Sitting across from him was a rather rude, obnoxious man who was aware of the great painter's accomplishments. The man tried several times to engage Picasso in conversation, but Picasso wanted only to relax and watch the scenery. Finally, in a loud voice the man blurted out, "There is nothing to this modern art stuff; I don't understand it at all; I think that the public is being duped!" Picasso could no longer ignore the fool. "Oh, you do?" said Picasso.

"Ya, I like art that is realistic," the man exclaimed.

"Well, what would you consider realistic art?" Picasso asked incredulously.

The man pulled out a picture of his wife from his wallet and handed it to Picasso. "Look at this picture of my wife. I like art that looks like the real thing."

Picasso examines the photo from various perspectives. "So, this picture is exactly how your wife looks, ha?" Picasso asked. "She certainly is small." And then turning the picture on edge he said, "And very thin too!"

Albert chuckles as he lights his pipe, thoroughly enjoying Niels' story.

NIELS: As a Buddhist friend once told me, we should never mistake reality for the phenomenon.

Albert walks across the room to examine a modern day illustrated version of the Schrödinger's cat experiment.

ALBERT: I love this. Where did you get it? I need one of these hanging on my wall.

The picture portrays a closed box containing a living cat, a bottle of cyanide, a rigged trigger device designed to smash the bottle if a quantum event occurs, and an unstable element which is undergoing radioactive decay. In this particular experiment the rate of decay of the radioactive element has been established, and it has been determined that after 10-minutes of elapsed time there is a 50-50 chance that a particle will be ejected. The trigger device is set so that it will function only for this 10-minute interval. If during this 10-minute interval there is a particle ejected from the radioactive element, it will trip the trigger, the bottle of poison will be smashed, and the cat will subsequently die. If no decay happens, then the trigger does not activate and the cat lives. The question is: What happens after an elapsed time of 10 minutes? (Figure 7.)

Figure 7.

NIELS: A friend gave it to me. It's fascinating. What is your assessment of Schrödinger's thought experiment?

ALBERT: Well, I think that he had a great point. He was really clever to so graphically show the absurdity of quantum effects on the macro scale. I don't think that we humans should be so egocentric to think that we are the only entities in the universe that can collapse the wave function, if that is what really happens. If consciousness really does play a role, then the cat could just as easily collapse the superposition of states into a single event. But, I think it more likely that consciousness is not involved at all. I think that the collapse has more to do with laws of thermodynamics. The universe extracts information and produces a small amount of entropy.

NIELS: Yes, on the ride home from your cabin the other day, I began to daydream about the whole objective meaning thing we had discussed. You said that meaning creates consciousness; consciousness does not create meaning. I really liked that thought. Then I started to wonder the obvious: What creates meaning? It occurred to me that we have a complementarity similar to my particle/wave complementarity.

ALBERT: Yes, your wave/particle complementarity idea was profound.

NIELS: The whole complementarity idea came about because of my study of Eastern philosophy. The Chinese called it the *ti* and the *t'ien*; two opposite, yet complementary, aspects of the same reality. I now believe that we must have an information/meaning complementarity.

Tibetan Buddhism offers the most succinct idea of the concept of complementarity. Buddhists maintain that nothing exists independently. We saw this idea in the monistic axiom of internal relations as well. Everything in the universe is interdependent. This is a departure from the classical description of physics in which the independent existence of matter is assumed. But quantum field theory has changed all of that. It is the relationship between things that is important. Physicists have come face-to-face with this idea in regard to particle physics. A quark cannot exist independently. It has no existence other than its relationship with other quarks inside the nucleus of the atom.

Al, as a classical physicist you would believe, and correct me if I'm wrong, that when we indirectly observe the particle-like aspect of the complementarity, the wave exists in reality even though the experimenter isn't able to observe the wave, and likewise, when we look for the wave-like aspect of the particle, the individual particle exists in reality even though we are unable to observe it. But this idea is inconsistent with quantum theory. Neither the wave nor the particle aspect of the duality exists when they are not observed. This is the departure from classical physics that quantum theory has provided.

I would like to state emphatically that this must also be true of the information/meaning complementarity. Neither information nor meaning has an independent existence in and of themselves. They must be considered together in the same way that your space/time complementarity must be considered together. That, I believe, is what John Wheeler has missed.

The eminent quantum physicist John Wheeler was a student of Niels Bohr. He is most famous for coining the term "black hole" and creating a thought experiment called the "Wheeler delayed choice experiment." He was the successor to Albert Einstein's chair at Princeton University.

He said in his early days he believed that atoms were the basic stuff of the universe; later on he thought that fields were the basic entity, but finally, he came to believe that information is the ultimate reality. All information is physical whether it is petroglyphs on rocks, ink on paper, the magnetic arrangements of atoms on a CD, the up/down spin of electrons, or the polarization of light.

Sure, all information has a physical aspect, as Wheeler and others have suggested, but it is the relationship between this physical attribute and meaning that gives it reality in any sense of the word. Ink on paper is just nonsensical scrawlings until one understands the alphabet. The electromagnetic arrangements of atoms on a CD are just chaotic manipulations until the CD is played and we hear the music of Mozart. Neither information nor meaning can exist independently.

ALBERT: Ok, go on.

NIELS: Earlier you were talking about how light was one of the most pure and basic aspects of nature. Information resides in light and our observation of that light gives it meaning. What causes the collapse of the wave? Possibly there is no collapse; rather, a transfer from one aspect of reality—information—to a reality of meaning. When we receive a certain wavelength of light that is reflected from the grass, for example, it impinges on our retina and the information is changed to meaning. But the cycle doesn't stop there. Once we perceive the meaning: the grass is green, for example, the meaning is stored by chemical and electrical means in the brain as information. The information has changed the brain; it has made new neural connections; the new neural connections establish new meaning which is then stored as information, which again changes the brain as the cycle continues. Then it occurred to me that light itself, traveling through space, undergoes a similar complementarity—the electromagnetic wave. The wave travels by changing from an electrical field to a magnetic field as it pulls itself along. Is it possible, I wondered, if the electromagnetic wave might itself be an information/meaning complementarity? This would provide an alternative explanation as to why a wave of light doesn't need a medium like the ether, and why quantum theory contends that the light has no independent existence from the time it leaves the source until the time it arrives at your retina. It is nothing but the *ti* and the *t'ien* as it travels through space.

ALBERT: I like it.

NIELS: The axiom of internal relations shows how meaning alone can affect reality. Change nature and you change an object's attributes; change the object's attributes and you change its meaning; change meaning and you change nature.

I now think I understand why you said that it wasn't necessary to bring "mind" into the discussion when we talked about the mind observing the brain in those OCD experiments. The information/meaning complementarity is self-referential so it is constantly changing. When we get new knowledge, such as obsessive thoughts are bad or smoking is bad, that information changes the meaning, and the changed meaning in turn changes the brain and the changed brain changes the meaning, ad infinitum. In this way mind and matter are the *ti* and the *t'ien*. They act together in an information/meaning complementarity. Neither mind nor matter has an independent existence. The complementarity can give us an alternative to the statistical wave that you find so repugnant.

ALBERT: I wish my old friend Max Born was still alive to hear you say that. Go ahead I'm listening.

NIELS: In regard to the quantum pea experiment, for example, Max would have believed that the pea resides as a statistical wave, and this has been the generally accepted dogma since the 1930s, but from our new understanding it would seem more appropriate to consider the statistical wave as an information/meaning complementarity. At the instant a particle is observed, the information/meaning complementarity changes to *information* which is physical. The information that we acquire by knowing its position changes the meaning of nature objectively. The objective change in nature in turn changes our individual meaning.

Al, you'll remember that out at your cabin we came to the conclusion that meaning was in fact objective, but meaning can be subjective when concerning individual experience. For example, the meaning of words depends upon our experiences. If I write a word in Chinese, it will definitely have a different meaning to you than to a Chinese speaking individual. But we must remember that we are a part of nature as well, so our subjective meaning is objective to the universe at large. So to expand the axiom of internal relations, we could state: change a thing in the brain and you change the universe; change a thing in the universe and you change the brain. Plato was correct after all.

ALBERT: Yes, I like that idea. Perhaps we should amend John Wheeler's proclamation that information is the basic entity of the universe, and instead say that the meaning/information complementarity is the basic entity of the universe. As David Bohm said, meaning is ubiquitous and non-local. While information must abide by the laws of thermodynamics and Relativity, meaning has no such constraints. Petroglyphs have no meaning until the information they encode is comprehended; words on paper are just scrawling until the alphabet is understood. In other words, information and meaning are both required for an expression of reality. On the quantum level the significance of this becomes apparent. That quantum pea had no physical existence under a particular cup until meaning was established through observation. After the observation its entire history was established. Meaning is not necessarily restricted to human consciousness. Meaning as we said earlier is a universal concept, and our consciousness experience is derived from the information/meaning complementarity, not the other way around. And I think that this is why we have two seemingly disparate types of consciousness: the consciousness produced by the brain and universal consciousness. Remember at the harbor I asked why we need all those pesky neurons in the brain if consciousness is universal. This solves that problem. If the meaning/information complementarity is the ultimate reality, then consciousness as we think of it is secondary. The meaning/information complementarity is an objective part of nature; our neurons are part of nature. The process of comparative actions "out there" is the same as the processes of comparative actions that take place in our brains.

NIELS: Ah. You used my idea of the information/meaning complementarity to its ultimate conclusion.

ALBERT: Plato thought that our subjective experience was a result of the external world of forms. He didn't believe in the notion of free will. So, if we believe in what David said about the implicate order, that nature is imbued with meaning at all levels of reality, then we can see that, in fact, the subjective aspect of meaning is really an objective part of reality. As we discussed out at my cabin, there isn't really a distinction between objective and subjective science after all. So the idea of free will becomes a non-question as well. What was thought to be a subjective decision to get up and go the refrigerator is nothing more than trillions of comparisons of meaning taking place in the brain. Your decision to act is a result of the most consistent meaning. Since the time of Kant, Locke, and Berkeley we had always believed that science must always be a subjective pursuit, but if meaning itself is objective, then in some respects we are indeed doing objective science.

NIELS: I don't want to interrupt your train of thought here, but I can think of an exception to the rule that information abides to all the laws of thermodynamics. It's true that energy can neither be created nor destroyed, but information can be created from nothing. You'll witness that when we do the experiments with light polarization.

ALBERT: Well, we'll see when the time comes. Let's apply our new idea of meaning and information to the Schrodinger cat experiment.

If the Geiger counter is tripped, information is extracted from the universe and that creates a small amount of entropy. That information is not known to us. It has not impacted our consciousness, but it is known to the universe at large. Meaning must be consistent. Meaning is primal. Therefore, if an atomic event occurs, and the vial of poison is broken, the cat dies whether or not we are looking. A meaningful event has been registered.

As we noted with the axiom of internal relations, all meaning is derived from comparisons of information. If I say that the planet Mercury is small, the statement is meaningless, unless I say that it is small compared to other planets in the solar system. Mercury is very large compared to a baseball for instance. So, let's go back to the original photon experiment, and use your hypothesis of the information/meaning complementarity. When the single photon is traveling, it carries information. If an observation is set up on the far side of the barrier at hole "A," for example, and information is extracted in the form of an observation, meaning is immediately established in the form of a history. We know which route the photon took. Meaning is non-local and instantaneous and this is why we get the apparent backwards-in-time illusion. Meaning is not restricted by relativity. When information is turned into meaning, a flash of consciousness occurs which is nothing more than the result of comparisons—the relationship between the information/meaning complementarity. The photon had no history between the time it was ejected from the source and its detection as hole "A." Likewise, the everyday reality we perceive with our minds is the result of millions upon millions of comparisons happening within our brains. We don't become aware of these comparisons until the information extracted becomes meaningful. Meaning is what we call our awareness and it is what creates our inner world of experience.

NIELS: I see. Remember when I told you that information always abides by the laws of thermodynamics, and I told you that information could be created from nothing in certain circumstances?

ALBERT: Yes.

NIELS: Well, let's go back to the pea under the tea cup analogy. You'll remember that the pea was in a superposition of states. A 50/50 possibility existed in finding it under cup A or cup B. You said this is an objective probability in that it is the same for anyone who looks. The two possibilities of finding it under cup A or cup B is a shared bit of information. It is a quantum bit or a qubit.

Now suppose you have a friend who looks under one of the two cups when you aren't around. The qubit collapses into a single bit of information for the person who peeked and for the universe at large. Meaning was created for the universe. But you still don't know which cup it is under when you arrive to see your friend. Now, you have a *subjective* probability of finding it under one cup or the other, because the pea is physically under one of the two cups, and not in a superposition of states. So in effect, there was another interesting thing that happened when the qubit collapsed. It actually created more information in the process.

ALBERT: How so?

NIELS: Well, for you, the person who is still ignorant of the situation there is a yes/no possibility that it is under cup A, and a yes/no possibility that it is under cup B; that's two bits of information, whereas before there was just one bit of information. Initially, the probability that you would find it under cup A or cup B was a quantum probability, a shared qubit, but now the situation has changed. It is now physically under one of the two cups. So meaning created information, and when you look under the cup you will extract personal information which in turn will create meaning. If, for example, you don't find it under cup A, you will automatically know it is under cup B, and vice versa. Information creates meaning and meaning creates information.

I agree with your idea that meaning creates the conscious experience of our inner world. When you learned which cup the pea was under, you had a meaningful experience in your brain which created a conscious experience. That conscious experience caused a physical change in your brain, and as we both know, all information has a physical representation. So, information was created in the process. Perhaps a new neural network was formed, of just a strengthening of a neural pathway, which in turn might have made another meaningful connection in turn causing another conscious experience. You get a cascade effect.

ALBERT: Interesting indeed. Our philosophies are moving closer together.

The phone rings and Albert uses the opportune moment to go over and raid Niels' stash of tobacco from a bowl sitting near the library of books. After he stuffs his pipe, he grabs a book, returns to his chair, and pretends to be engaged in some reading. But Niels is not fooled. After hanging up the phone he confronts the embarrassed Einstein.

NIELS: I thought your doctor said that you weren't supposed to smoke anymore.

ALBERT: No, his exact words were that I should not *buy* anymore tobacco. I have followed his advice to the letter.

Albert displays a sheepish grin as he draws a match, strikes it, and lights his pipe.

NIELS: That was John on the phone. He should be here any minute.

ALBERT: Really!

NIELS: Ya, and don't forget the three of us have that presentation this weekend. Are you prepared?

ALBERT: No, not at all. You?

NIELS: Nope, me either.

Albert, Niels, and John have agreed to a "meet-the-students" presentation in which they will discuss the upcoming Bell project and answer question from the faculty and student body. The symposium raised so much interest that Princeton officials had considered holding it in the basketball arena to accommodate the large crowd; however, the three presenters refused such a large forum. Instead, they decided to hold it in an auditorium that seats around four thousand people. Many interested participants have been turned away as a result.

Niels hears a knock at the door.

NIELS: Come in, come in.

JOHN: My god, it is such an honor to meet you both.

ALBERT: We have been waiting in great anticipation. How are you?

Albert, Niels, and John are at last all together and they take the opportunity to get to know more about each other. John is completely awe struck in the company of these two mythic figures, and at first he finds it difficult to even speak, and though he never experienced the fame of Albert and Niels, during his lifetime, he certainly is their intellectual equal. The three exchange small talk for about half an hour, and then Rashid and Risa come over to join them. Risa has come with pen and notebook in hand to do a feature story on John for the school newspaper.

RISA: Doctor Bell, so nice to meet you. Would you mind if I ask some questions for the paper? We're doing a feature article about your life.

JOHN: Sure, that would be okay if the others don't mind. Please, just call me John.

NIELS: That's fine with me. I'm interested to hear your story as well.

RISA: Wonderful; if either of you two have any comments please feel free to join in. Then I'll get started: When were you born and what was it like growing up in Ireland as a boy?

JOHN: I was born in Belfast in 1928, right at the pinnacle of the quantum revolution. My dad's name was John as well, so most people called me by my middle name, Stewart. I had an older sister named, Ruby, and two younger brothers, David, and Robert. Of the four of us, I was the only one to stay in school after the age of fourteen. We were very poor and universal education did not exist in Ireland at that time, so going to school was a luxury. But, because I got encouragement from my mom and my grades were good, we were able to raise enough money for me to go to the Belfast Technical High School. Their curriculum qualified me to attend the university.

RISA: Where did you go to college?

JOHN: I didn't go to college immediately after high school because I couldn't afford it, so I signed on as a technician at the physics department at Queen's University Belfast. Some of my instructors were kind enough to let me sit in on some of their lectures; by 1945 I was able to enroll as a student and within three years I had attained a degree in experimental physics.

RISA: Did you continue on to get your Ph.D. at Queen's?

JOHN: No, after getting my degree in physics, I stayed around for another year and got a second undergraduate degree in mathematical physics, but by that time the money ran out, so I got a job at the Atomic Research Establishment at Harwell in the United Kingdom. After that I moved to the accelerator design group in Malvern. It was there that I met my wife, Mary Ross, who also had degrees in mathematics and physics. We married in 1952.

Albert thinks back to the time when he met Mileva at the polytechnic and how his attraction to Mileva was both physical and intellectual.

JOHN: Shortly after we got married, Mary and I co-authored several papers on the meaning of quantum theory, and in 1960 we moved to CERN where I continued my work. I believe I published something like 80 papers on high-energy physics, but I never lost sight of my concern about the meaning of quantum theory. The whole business made me very uneasy, so while on Sabbatical from CERN, I decided to take another look at Albert's EPR paper that was so important to his defense of objective reality.

ALBERT: Ya, I appreciate your effort.

RISA: What happened after you published your famous theorem of inequality?

JOHN: Almost nothing. My proof went largely unnoticed for years.

ALBERT: I know the feeling.

JOHN: It wasn't until the last few years of my former life that I began to get credit for my theorem. I eventually received the Hughes Medal of the Royal Society, the Dirac Medal of the Institute of Physics, and the Heineman Prize of the American Physical Society. I was also nominated for the Nobel Prize, but I never lived to receive it.

NIELS: I'm sure you would have won. No question in my mind.

JOHN. It wasn't until my death in 1990, that the accolades really started to come in. Now everyone seems to be interested in my theorem.

NIELS: Yes. I have known several physicists who have called your theorem of inequality the greatest scientific achievement of all time. One thing is certain. Even if quantum theory is someday replaced, or at least modified, your theorem will stand.

JOHN: Yes, because it has now been proven experimentally. I never expected that to happen during my lifetime, but technology moves on.

NIELS: The first tests of your theorem were performed in Berkeley, California, but the technology wasn't precise enough to rule out other factors that might have influenced the results. It wasn't until 1982 that your theorem was proven definitively by Alan Aspect and his colleagues at the Institute of Optics at the University of Paris at Orsay. That was certainly a triumph for experimental physics and your theorem of inequality.

JOHN: Yes, they were able to verify my theorem to a distance of a few meters. Nicolus Gisin and his group were able to verify the results up to seven miles, and I think that is where it stands now. I have little doubt that my theorem will hold for any distance.

NIELS: I guess that's why we are here. In just a few weeks we'll know for sure.

Risa continues to ask John questions and after a bit Rashid and Risa excuse themselves leaving the three prominent scientists to continue to get acquainted. For the next hour the three converse about the upcoming experiment and John's work out in Berkeley. John then says goodbye and returns to his office just across the hall.

The three visit each other's offices regularly over the week to prepare for the upcoming symposium and often Rashid and Risa stop in to talk physics and take notes for articles in the school paper.

Finally, the big day arrives and Albert, Niels, and John depart Jones Hall in the company of Rashid, Risa, and several security guards. They had been offered rides in a limousine, but they did not want the pomp and ceremony. As they proceed down the side walk they are mobbed by the press and well-wishers seeking autographs. The sound of cameras taking photographs fills the air. One persistent student walks along side and attempts to engage Albert in conversation.

STUDENT: Albert! Do you think time travel is possible? Can we go back in time?

ALBERT: It is theoretically possible to travel into someone else's future. I'm not so sure we can go back in time, however.

Albert turns to Niels and in a low voice mutters to Niels:

ALBERT: I feel a bit like Steven Hawkins when he is asked about the Schrödinger's cat experiment.

NIELS: How so?

ALBERT: Hawkins use to say that whenever someone brings up that experiment he wants to get his gun. Nothing bores me more than the time travel thing, and yet that is what everyone wants to know about for some reason. Too many science fiction movies I suppose.

A young female student catches up to the group and fires a question at Albert:

STUDENT: Sir, do you think that the average person will ever be able to understand your theory of relativity?

ALBERT: It's not so difficult really.

STUDENT: You're so modest.

ALBERT: No, I'm just stating the truth. I'll bet you can understand it if you knew what relativity theory really says.

STUDENT: I doubt it, but can you tell me in a nutshell?

ALBERT: Relativity theory states that the laws of physics are the same for all observers in the same frame of reference.

STUDENT: That's it? It doesn't state that all things are relative.

ALBERT: Not at all.

STUDENT: Wow!

The three proceed toward the assembly hall where they will speak. Just outside the hall a band, taking advantage of the large gathering, is playing rock-and-roll music. A sporting event has just ended in a nearby venue, and hundreds of students have spilled out into the street to listen to the band. Many have been consuming alcohol, and the air is thick with the smell of marijuana. One of the band members sees the entourage approaching and invites them to come on stage. Niels and John have no intention of going on stage, but Albert has never seen a microphone he didn't like.

ALBERT: What do you say, you guys? Let's go.

JOHN: Are you nuts! Most of these kids don't even know who we are.

NIELS: They're all drunk.

ALBERT: Come on. Don't be such stick-in-the-muds. Besides we're early. Let's give these students their money's worth. Some of them have paid good money to get in to hear us talk.

Incredulously, Albert climbs the steps and walks on stage. The others follow because at this point they really have no other choice. The raucous crowd cheers wildly, followed by whistles and cat calls. All sense of protocol and decorum dissolve as the irreverent crowd pushes toward the stage. Albert makes his way to the microphone while the others hang back, totally uncomfortable with the situation. Albert is the most recognizable and most of the questions go to him.

STUDENT: Hey Albert! How do you like the cafeteria food here?

ALBERT: The food is terrible and the portions are too small.

The crowd hoots with approval.

STUDENT: Albert, how did your hair get that way?

ALBERT: I don't know really, I woke up one morning and there it was.

Laughter

STUDENT: What do you like best about Princeton?

ALBERT: The co-eds.

Albert is on a roll and doesn't even think about relinquishing the microphone to the other two.

STUDENT: Are you religious?

ALBERT: I take the same view on religion as Voltaire.

STUDENT: What's that?

ALBERT: I think that religion was created when a charlatan met a fool.

The crowd explodes with laughter, now emboldened to ask more questions.

ALBERT: How do you like your accommodations here?

ALBERT: Not too bad, but I'm not getting much sleep.

STUDENT: How come?

ALBERT: Well, last night I was awakened at 3:30 by this pounding on my door. A young woman was pounding and pounding on my door and I was afraid she'd wake the neighbors.

STUDENT: What did you do?

ALBERT: I finally had to get up and let her out!

Albert continues to delight the crowd with one-liners, until he feels Niels tugging on his sleeve. Niels tells Albert to wrap it up as the time is nearing to go inside the venue for the symposium. Albert turns back to the microphone to bid everyone goodbye.

ALBERT: Well, you've been a great crowd. We hope to see some of you inside. Party on!

The crowd bursts into applause having gained much respect for Albert who they now consider one of them. Niels and John are embarrassed and exit the stage as fast as possible.

NIELS: Well come on down Shecky Greene. I think you missed your true calling, Al.

ALBERT: What do you mean?

NIELS: That was a pretty good stand-up act.

The three give their presentations at the assembly hall to a respectful and appreciative crowd. The question and answer period lasted into the night as students clamor to the fore to get a chance to ask questions. Finally, the meeting was called to a close and the three are once again ushered back to their offices without incident.

The next two weeks finds the three researchers in intense preparation for the upcoming Bell Project experiment. Data from the orbiting probes will be beamed to the Jet Propulsion Laboratory in Pasadena, California, where it will be decoded and sent in a comprehensible form to Princeton university physics laboratory, The First Campus Center. There, the data will be analyzed by a team of scientists which include Albert, Niels, and John.

The Bell Project

THE BELL PROJECT named after John Stewart Bell is an experiment designed to test Bell's theorem of inequality, which demonstrates the proposition that reality is non-local. Previous experiments in the 1980s and 1990s have proven the validity of Bell's theorem of inequality to a distance of about seven miles, but this will be the first experiment to validate Bell's theorem out to a distance of millions of miles. Most are convinced of the validity of Bell's theorem, but some skeptics remain, including Albert.

NASA has launched three probes into deep space. Two of the three probes have been put into *Lagrangian points.* Italian-French mathematician Joseph Louis Lagrange discovered in 1772 that for any two orbiting bodies in space, there are five discrete positions in space that a third body of insignificant mass can be placed, such that they will remain stationary relative to the two large bodies. Two of the probes designated as Bell 1 and Bell 2 were launched in 2028 into two of the Lagrangian points between the Sun and the Earth. The two probes are several million miles from the Earth, and the two probes are separated from each other by several million miles. The third probe called Photos was launched away from the Sun in a direction toward the outer planets. Photos could not be placed in a Lagrangian point because that location would have put it equal distance from the other two probes. For this particular experiment Photos had to be placed five light minutes closer to Bell 1 than Bell 2. As a result, the relative stationary positioning for Photos requires precision gyroscopes, laser equipment, and fuel to power the retro rockets to keep it stable at the appropriate distances relative to the Bell 1 and Bell 2. All three probes are now in their respective positions and have been given the go-ahead by NASA to carry out the experiments.

The polarizers on board Bell 1 and Bell 2 will have their angles set in advance of each part of the experiment through radio signals sent from NASA. After the results of each set of measurements have been logged, the probes beam down the results to NASA's Jet Propulsion Laboratory in Pasadena, California. The results are then decoded and sent via the internet to the researchers at Princeton.

The enormous space/time separation of the probes is required because of the extremely fast velocity that light propagates through space. The vast distances will make certain that no communication can take place between the probes in violation of Relativity, and once and for all remove any doubt of the validity of Bell's theorem of inequality. In addition, the large separation of the probes will allow scientists to confirm or deny the validity of Bell's theorem of inequality to infinity.

In this particular experimental set-up, correlated photons produced from excited atoms will be shot in opposite directions from a photon gun aboard Photos. One photon will be shot toward Bell 1, and its correlated twin photon will be shot toward Bell 2. Photons that are in a twin state of parallel polarization have a unique property in that they have the same angle of polarization for all angles measured. Neither photon has an angle of polarization until it is measured however. The only thing we know for sure is that whatever angle of polarization one photon has when measured, its twin will have the same polarization when it is measured. For example, if the first photon to arrive at the Bell 1 probe is measured at a "vertical" angle of polarization, its twin arriving about five minutes later at Bell 2 will also have a "vertical" polarization. Likewise, if the first photon is measure to be "horizontally" polarized, its twin will have a "horizontal" polarization angle as well. Previous experiments with correlated photons have shown conclusively that identical angles of polarization occur in every instance. John, Albert, and Niels all agree on this fact of quantum theory.

Incidentally, the terms "horizontal" and "vertical" are used for convenience and do not mean that the photons are actually oriented to some preconceived horizontal or vertical plane in space. A horizontal or vertical measurement is completely relative and defined by the measuring apparatus itself.

The disagreement between Niels and Albert hinges on two key points: Is nature objective and determinate or subjective and indeterminate. Bohr adheres to the philosophy of indeterminism and chance. He believes that the photon has no angle of polarization until it is measured, and he believes that for each measurement the photon has a 50/50 chance of either being oriented to the angle set by the polarizer, or perpendicular to the angle of the polarizer setting. Einstein on the other hand is an objective determinist, believing that the angle of polarization is determined for both photons at the outset. This experiment is designed to settle these disputes. (Refer to polarization experiment page 10.)

We remember from our earlier discussion of this experiment that polarization is the simplest dynamic attribute as it can take only two possible values. For each direction a single photon has only two options: either it is entirely polarized in the direction measured or entirely polarized at right angles to the measurement (Herbert, 136). If the photon passes through the polarizer, meaning that it is parallel to the optical axis, the photon will be recorded as an H, meaning a hit. On the other hand if the photon is measured to be perpendicular or horizontal to the optic axis, the photon will take another path toward a second counter and be marked with an M for a miss.

The long anticipated day has arrived, and the experimenters have taken their positions at The *First Campus Center,* where about twenty-five scientists and engineers have gathered to sort out the data arriving from the Jet Propulsion Laboratory.

In the first part of the test, both polarizers, one on Bell 1 and the other on Bell 2, are set at the same angle of orientation relative to each other. We'll call this setting "vertical." At the specified time, the space probe Photos begins to fire single correlated pairs of photons. One photon is shot toward the Bell 1 probe, and its twin photon is shot toward the Bell 2 probe. After traveling through space for several minutes, the first photon of the pair arrives at the Bell 1 detectors and its measurement is recorded. About five minutes later the second photon of the pair arrives at Bell 2 where it is recorded. After firing hundreds of individual pairs of correlated photons at their respective targets the results are logged and counted and the results are beamed to the Jet Propulsion Laboratory and then on to Princeton for review. The results are as follows:

Bell 1 >>>>>>> H–M–M–M–H–M–H–H–M–H
Bell 2 >>>>>>> H–M–M–M–H–M–H–H–M–H

With both polarizers set to the same angle a 100% correlation was found, confirming that both photons, upon measurement, have the same angle of polarization. If the polarizer on Bell 1 registered a "hit," then the polarizer on Bell 2 registered a "hit." If the polarizer on the Bell 1 registered a "miss," then the polarizer on Bell 2 registered a "miss."

Nothing is surprising here. This confirms the well- known fact that paired-correlated photons always have the same angle of polarization when they are measured. The researchers review the data, and all are pleased that there are no glitches or surprises.

In the next part of the experiment the polarizers are off-set 90 degrees from one another. After firing hundreds of pairs of photons the results are as follows:

Bell 1 >>>>>> M–H–M–M–H–M–H–H–M–H
Bell 2 >>>>>> H–M–H–H–M–H–M–M–H–M

In this case a 100% anti-correlation was found upon comparing the results at the two locations. This demonstrates once again that the photons have the same angle of polarization after they are measured. As expected, if a photon is measured to be parallel to the optical axis it will go straight through the first detector on Bell 1, go into the "up chamber", and recorded as a hit. Its twin, which has the same polarization, will be deflected into the "down chamber", because the detector is offset by 90 degrees. This agrees with the first part of the experiment: The correlated pairs of photons always have the same polarization angle when measured.

In both sets of experiments there exists a 50/50 chance of getting either a hit or a miss when the first photon is measured, demonstrating that the laws of probability hold and that the photon angle is completely random.

This concludes the first part of the test. The team of scientists and engineers take a break for the day and will return tomorrow for the second part of the tests. Niels, Albert, and John meet over coffee at a corner restaurant to discuss the first set of results.

JOHN: Well, Al, so far quantum theory holds wouldn't you agree? As you can see there is a 100% correspondence between the two polarizers. It's the relative angle between the two polarizers that is important here—not some objective angle of polarization that has been established from the start. No signal can pass from one photon to the other because they are flying apart at the speed of light, so assuming Relativity is correct we can rule out any kind of classical communication between the photons. I think we can say that neither photon has any angle of polarization until it is measured.

ALBERT: I'm not ready to say that the angle of polarization is random, nor that the polarization is undetermined at the source.

NIELS: Al, if the angle of polarization is determined at the source, as you suspect, there would not be a 50/50 chance of a hit or a miss. As you can see, the photon is either oriented to the exact angle the experimenter chooses or exactly perpendicular to the angle that is chosen. Never is it at some in-between angle to the chosen setting of the detector. If the angle was determined at the source, this would not be the case.

ALBERT: You can't conclude from this part of the experiment that the polarization is either at the angle measured or 90 degrees from the angle measured.

NIELS: Why not?

ALBERT: Even if I agree that the photon travels as a statistical wave, or as an information/meaning complementarity, we cannot tell from this experiment that the photon's polarization is not determined at the source. Suppose that the polarization angle is determined at the source of departure as I contend, and the polarization angle at Bell 1 is set to a vertical angle; let's call it 0 degrees. And suppose that the Bell 2 polarizer is set to a horizontal position, relative to Bell 1's polarizer. We'll call it 90 degrees. If the photon is a real wave, or a statistical wave for that matter, and it begins its journey polarized to any angle between 0 degrees and 45 degrees positive or negative from vertical, it will pass through the polarizer and register a hit. That is because any part of the wave that hits the polarizer will set off the detector as if the whole photon passed through. Its twin, which has the same angle of polarization, will then register a miss at Bell 2. On the other hand, if the photon leaves its source at Photos with an angle of 45 degrees or greater, no part of it will go through the vertical polarizer on Bell 1 and it will register a miss, while its twin will register a hit when it arrives at Bell 2. So that is the reason you think it is an either/or proposition. You're not taking into account the in-between angles. Any photon wave that is within 45 degrees positive or negative will register a hit if the polarizer is set vertically, and any photon with a polarization of more than 45 degrees will register a miss when the polarizer is set vertically.

JOHN: I think that we'll be able to put those concerns to rest when the second part of the experiment is conducted.

ALBERT: We shall see, but I'm also concerned about the randomness aspect of this experiment. I don't believe in the idea of "laws of probability." I don't deny chance. I have a chance of getting hit by a car. Princeton has a chance of winning its basketball game. A chance exists that one of the space probes will malfunction, and the experiment will be ruined, but no *law* of probability exists. Friedrich von Schiller once stated that "There is no such thing as chance, and what we regard as blind circumstance actually stems from the deepest source of all." Probability is a relative term. Probability only makes sense in a casual world of space, time and dimensionality. I contend that it is impossible to distinguish chance from determinism. Randomness is merely our inability to perceive order in chaos, to see patterns in nature. Remember the example we talked about with the nuclear decay problem. You thought it odd that a radioactive substance has a half-life. But we have seen that this is really an issue of statistics, not probability. A statistical outcome describes what happens after the fact. It is a compilation of events in time and space. Statistical outcomes require a history. We could never extrapolate the half-life of an element in real time. It must be calculated in retrospect after looking at the results over time. The very fact that each and every sample of a particular element has the same half-life is a testament to the fact that is not random but predetermined. The statistical likelihood of a couple having ten girls in a row and no boys is one chance in 1024, but the likelihood of the next child being either a boy or a girl is still 50/50. As we talked about before, classical randomness is subjective, and quantum randomness is objective. Remember, we determined that subjectivity is an objective part of the universe. The laws of the universe are objective and deterministic; therefore, randomness is objective and deterministic as well. It is just that we don't know the algorithm. Maybe someday we will. Also, when you talk about the "laws of probability" you are by definition making it objective. Laws are the same for everyone. The moment you establish probability to be law, it can no longer be random.

NIELS: The semantics of this issue is a problem, I'll admit. And there is no agreement on just what probability is. Some philosophers believe that chance is nothing real and others think that it is the ultimate reality.

JOHN: I think that we can all agree that whatever polarization one photon has, its twin will have the same polarization. I think that the reason they are correlated is due to the law of indistinguishability as Satyendra Nath Bose and you, Albert, figured out long ago. You and Bose established that when one attempts to figure out the pressures for confined gases, one does not need to multiply the pressure of each particle by every other particle to establish the pressure of the whole. One only needs to count the particles once. The idea here is that if one cannot distinguish one particle from another, they are for all practical purposes the same particle. So, I think that we can use the law of indistinguishability if we assume that neither photon has the attribute of polarization before it leaves the source.

ALBERT: Ok, explain.

JOHN: The law of indistinguishability offers an explanation as to why Niels and I believe that the photon need not have a specific polarization as it leaves its source. Since the photons have no attributes, there can be no difference between the two photons, so we can say they are indistinguishable and, therefore, identical particles in all respects. Since they are mathematically the same we can say they share a single qubit of information. If we extract information from one photon, we automatically have information about the other, since they share a qubit of information. Mathematically speaking and physically speaking, the two particles are really only one. They are non-local particles. So, after a measurement is made, information and meaning is extracted for both photons, and the meaning is transmitted instantaneously.

NIELS: It was you, Al, who once said that the only incomprehensible thing about the universe is that it is comprehensible. What else would better explain that statement than the fact that nature has meaning? Meaning is what the intellect strives to attain, and meaning is what constitutes nature. This is why we are able to comprehend the universe. Your friend David Bohm was correct when he talked about the implicate order. It is only because meaning is enfolded into every layer of nature's complexity that we are able to gather information through our senses and make a meaningful picture. Consciousness is a result of this process.

ALBERT: I agree.

The next morning Albert, Niels, and John, gather at the First Campus Center to participate in the second and final part of the experiment. If all goes as planned, John's theorem of inequality should be confirmed.

The essence of John Bell's theorem is put to the test in the second part of the experiment. NASA sends radio signals to Bell 1 to set the polarizer to an angle of 30 degree clockwise of vertical and a radio signal to Bell 2 to set its polarizer to a vertical position or 0 degrees for a separation angle of 30 degrees between the two polarizers.

One hundred pairs of correlated photons are fired from Bell 1 toward the two destinations. The results show that about one out of every four photons is a mismatch.

In the next phase of the experiment, the situation is reversed. This time the polarizer aboard Bell 1 is set to 0 degrees and the polarizer aboard Bell 2 is set to 30 degrees counterclockwise of vertical for a separation once again of 30 degrees between the two polarizers. As expected, after firing 100 pairs of photons in a twin state of polarization the statistical result show about one mismatch out of every four photon pairs fired. Shown below is a representative sample from one of the of the two tests results.

Bell 1 >>> H–H–M–M–M–H–H–M–H–M–H–M–H–H–H–M–H–M–M–H
Bell 2 >>> H–H–M–H–M–H–M–M–H–M–M–M–H–H–M–H–H–M–M–M
^ ^ ^ ^ ^ ^

The arrows indicate the mismatches.

One might wonder why there is, statistically, one mismatch out of every four photons fired when it has been shown that the photon is either polarized to the angle measured, or 90 degrees from the angle measured. In this case the angle of separation is not 90 degrees, but 30 degrees. Since the photon travels as a statistical wave occasional mismatches occur when there is a separation angle between 90 degrees and 0 degrees.

In the final part of the experiment, the polarizers on Bell 1 are set to a clockwise position of 30 degrees of vertical, and the Bell 2 polarizer is set to 30 degrees counterclockwise of vertical, for a separation angle of 60 degrees. Bell's theorem of inequality says that if the angle of polarization is determined at the source the number of mismatches cannot exceed two mismatches out of every four photons fired. Let's see what happens in the experiment after 100 pairs of photons are fired. The results of the experiment are sent back to NASA and read as follows:

Bell1>>>M–M–H–M–M–H–H–M–H–M–H–M–H–H–M–H–M–M–H–M–H–M
Bell2>>>H–M–M–H–M–H–M–H–M–M–H–M–M–M–H–M–H–M–M–H–H–H
^ ^ ^ ^ ^ ^ ^ ^ ^ ^ ^ ^ ^ ^

The arrows show the mismatches.

Surprisingly, the experimental data shows about three mismatches out of 4! Bell's theorem of inequality is violated, which means that nature is non-local as John predicted.

Albert, John, and Niels are given news of the results. (*Figure 8*)

NIELS: Congratulations John. Well done.

JOHN: Thank you. To quote Albert when he was notified that his theory of General Relativity had been experimentally confirmed: I would have felt

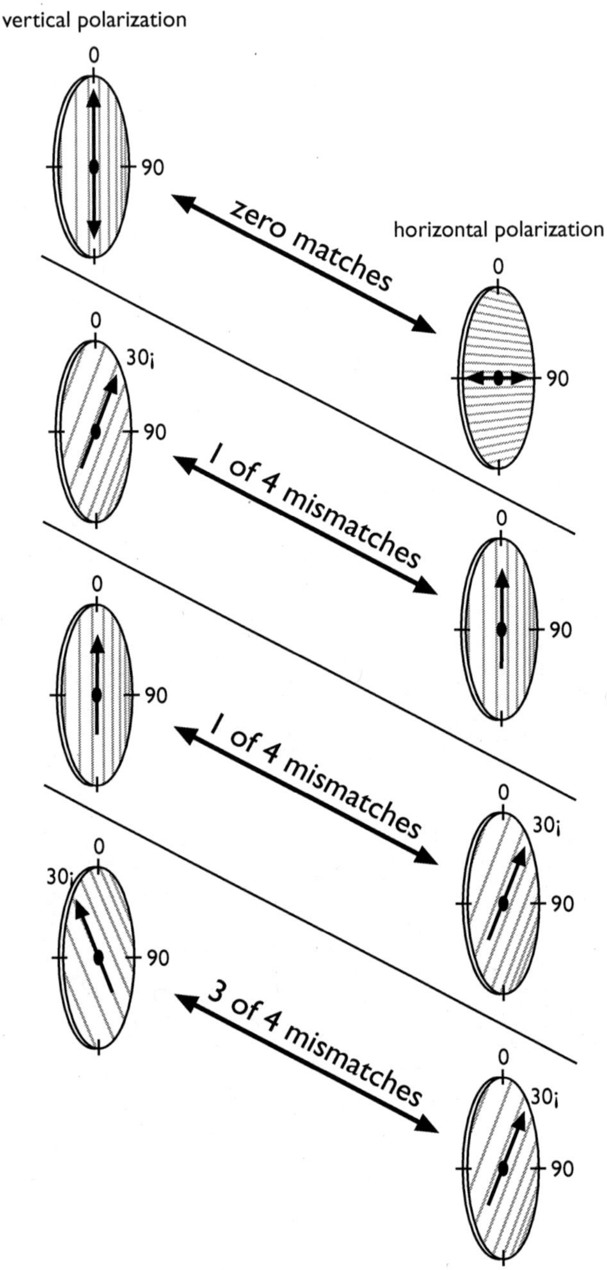
vertical polarization
0
90
zero matches
horizontal polarization
0
90
0
30°
90
1 of 4 mismatches
0
90
0
90
1 of 4 mismatches
0
30°
90
0
30°
90
3 of 4 mismatches
0
30°
90

Figure 8.

sorry for the poor lord if the results had been otherwise. This is exactly as expected.

NIELS: Let me see if I interpret these results correctly. And I'm just going to think out loud here. When the experimenters had an angle of separation of 30 degrees and hundreds of pairs of photons were fired, there was about one mismatch out of every four pairs fired. This result confirms standard quantum theory that the photon travels as a statistical wave. It is also a confirmation of Maul's law which states that the cosine squared of the difference between the angles will give the probability of hits or misses.

John, what is the cosine of 30 degrees?

JOHN: Ah, let's see. It would be .8660. So square that and we get .75 rounded off.

NIELS: Yes, .75 or 3/4. We got three out of four hits which correspond perfectly with the experimental results and Maul's law.

NIELS: And the cosine of 60 degrees is .5. Square that and we get .25 or 1/4. When we had a separation of 60 degrees Maul's law says we should get one match out of four (three mismatches out of four) and that is exactly what the experiment proved.

JOHN: If the polarizer on Bell 2 is set to 30 degrees, either clockwise or counterclockwise from vertical and the polarizer on Bell 1 is set to 0 degrees there should be about one mismatch out of four, or in terms of Maul's law, three hits out of four and that is what the results showed. Likewise, if the polarizer on Bell 1 is set to 30 degrees from vertical and the polarizer on Bell 2 is set to 0 degrees, there should also be one mismatch out of four, and once again this corresponds with the results of the experiment.

But now here is the important part of the experiment. If the polarizer on Bell 2 is set to 30 degrees clockwise from vertical and the polarizer on Bell 1 is set to 30 degrees counterclockwise from vertical my theorem of inequality states that: if the photon already has an angle of polarization before it is measured then there cannot be more than two mismatches out of four. The results show three or more mismatches out of every four pairs fired, so the inequality is violated as my theorem states. Therefore, the photon cannot have an angle polarization before it is measured.

ALBERT: Explain to me very clearly your reasoning on this.

JOHN: Ok, in the experiment the correlated photons are shot toward the two destinations. We saw that when the polarizer on Bell 1was offset 30 degrees from the polarizer on Bell 2, the results showed one mismatch out of four. We got the same results when the situation was reversed. Suppose just for a moment that the photon's polarization is determined at the source as you claim. The photons leave their source at Photos and head toward their destinations. The photon arrives first at Bell 1 and is measured and recorded. Minutes later the twin photon arrives at Bell 2 and it is measured and recorded. As we stated, there is no causal relationship between the photons since they are racing apart at the speed of light.

We know that if the polarizer is offset by 30 degrees from "vertical" relative to the source at Photos we get one out of four mismatches. If the situation is reversed and we off-set the polarizer on Bell 2, 30 degrees from "vertical" relative to the source we also get one mismatch out of four. At this point, as you reminded us Al, we still cannot determine if the polarization is determined at the source. But now if we off-set the Bell 1 polarizer by 30 degrees clockwise in relation to the source, and off-set the Bell 2 polarizer by 30 degrees counterclockwise to the source we have a separation angle of 60 degrees between the two polarizers. If the photons are polarized at the source we should get two or fewer mismatches between the two polarizers. It is a simple addition problem. We should get one mismatch out of four as a result of the 30 degree off-set from the Bell 1 polarizer and another one mismatch out of four as a result of the 30 degree off-set from the Bell 2 polarizer. But as the experiment showed there are over three mismatches out of four which is consistent with Maul's law of a 60 degree off set.

It shouldn't matter to the photon arriving at the Bell 1 polarizer what angle the Bell 2 polarizer is set if there is no communication between the two photons. The only conclusion here is that it is the separation of the angle between the two polarizers that is important. This emphatically demonstrates that it is the measurement itself that determines polarization. The photons cannot have a polarization angle before a measurement is taken.

ALBERT: Ok, I understand the reasoning and your logic is sound. This experiment certainly confirms your theorem, John, but I'm still not ready to believe that it has all that much to do with the measuring device or the observer. A few days back Niels and I were talking about how mathematics is the most meaningful and fundamental aspect of reality. Specifically, we were talking about how Kepler and Newton each came up with an explanation of why the planets orbit as they do. Newton's idea has to do with forces, but Kepler's idea was purely mathematical. It turned out that Newton's idea was replaced by my relativity explanation, yet both Newton's idea of gravity and Kepler's mathematical description are equally valid in predicting the velocity needed to place an object in orbit for example. As Niels is fond of saying, physics concerns only what we can say about nature, not about how nature "is." We have no better understanding of what gravity is than the ancient Greeks. As positivists all we can hope for is a mathematical description of reality. Do you agree with that?

JOHN: Sure.

ALBERT: The question here is: Are the laws that Newton and Kepler contrived subjective or objective? Was Newton's mathematical formulation, which states that the force between two bodies diminishes with the inverse square of the distance between them, invention or discovery? Did he simply uncover the mathematics that is inherent in nature? In other words, does nature have math as its operating system?

Let's look at this experiment another way. I will now agree with you that the paired photons share one bit of information between them. As they travel they are indistinguishable because they have no inherent attributes. But it isn't the measuring device that gives the photon its dynamic attribute of polarization. It is Maul's law in and of itself.

Once the first photon is measured, the second photon must conform to Maul's law in order to conserve meaning. Maul's law is an objective part of the universe. Meaning is the most fundamental aspect of nature and mathematics is her language.

Never the one to disguise his feelings John slaps his forehead in exasperation at Albert's intransience and almost simultaneously Niels, who is sitting at his desk, covers his face as his head falls limply into his hands. One can barely hear him muttering.....Albert...Dalbert...Schmelbert! Niels is frustrated at Albert's stubbornness, but deep down he suspects that Albert might be correct. John tries to salvage the good fortune the day has brought and speaks:

JOHN: Well, it's been a long day, why don't we discuss this tomorrow. One thing is certain, this has been a momentous few days, and the events will be recorded for posterity. We now know unequivocally that my theorem is as sound as the bedrock on which we stand.

ALBERT: Yes, and for that you deserve congratulations. I think we should all go out and celebrate. What do you think Niels? Dinner and drinks are on me.

NIELS: Alright. I don't want to turn down this once in a lifetime offer. Let's go I'm starved.

The three go out to dinner at a nearby restaurant that they have often frequented. All disagreements are put aside as they enjoy an evening of food, drink, and laughter. Many friends, dignitaries, and acquaintances stop by to congratulate the trio and to take part in the celebration which lasts into the late hours of the night.

The next morning news hits the press about the confirmation of Bell's theorem of inequality, and the journalists clamor for interviews, but the three are having none of it and refuse all request. That evening they gather in Albert's office with Rashid and Risa as their only guests. Albert, Niels, and John are all in a nostalgic and reconciliatory mood because they realize their time is limited. Rashid begins to ask questions as Risa takes notes to record the conversations.

RASHID: Well, what have we learned to this point?

NIELS: Our time has been very productive over the past months. I believe the most important thing Albert and I have concluded is that the dualist notion of subjectivity and objectivity is no longer a valid picture of reality. We can no longer say that this is subjective and that is objective because the distinction between the two no longer exists in a non-local universe.

Another huge leap we made in our discussions is that the dualism of determinacy and indeterminacy is beginning to crumble. We still have some significant differences here, but Al has come very close to demonstrating to me that such things as nuclear decay can be seen as a determinate process. And, because of our new understanding that meaning is an objective part of the universe and that mind and matter are in reality an information/meaning complementarity we have concluded that the debate as to whether or not we have free will is meaningless. We have no more free will than the universe at large, and as a result, we have come full circle to embrace the Platonic view of the world.

ALBERT: Yes, that's right. And don't forget about the quantum-eraser experiment: A conscious observer is not required to collapse the wave function or the information/meaning complementarity as it will now be known. It seems that the collapse is a result of the universe extracting information and meaning. We have also concluded that mathematics is a fundamental aspect of the universe, and that John's theorem of inequality clearly demonstrates that nature is non-local. I believe that in the past few months our philosophical debates have eclipsed the great Solvay debates of 1927 and 1930 for their significance. It's been a great time. I'm glad we could all get together once again.

NIELS: I agree, Albert, John, it has been an unforgettable experience.

JOHN: Life the second time around is always sweeter.

Rashid and Risa continue to ask for more explanations as to how the three have arrived at their conclusions and after about an hour the two decide to pardon themselves as it is getting a bit late. When Rashid and Risa return to their office, which has doubled as an apartment over the past few months, Rashid grabs Risa's manuscript from her desk, climbs into the make-shift futon bed, and begins to read himself to sleep.

Meanwhile, Albert is standing in front of his office window looking out and he begins to reminisce about old times as the twilight slowly overtakes the city.

ALBERT: Ah, Princeton has such a beautiful campus. Ya know Elsa and I toured many campuses throughout America, including Berkeley. They offered me a lot of money, but when I saw this campus I knew I had found a new home. The architecture reminds me of the great universities of Europe.

JOHN: When did you settle here?

ALBERT: We came here in October of 1933. I had always planned to go back to Europe, but because of the war it never worked out. I was here twenty-two years and I have no regrets about not getting back to Europe. It was a very tranquil time. My stepdaughter Margot and I would take long walks in the woods; not talking much; just enjoying being out in nature.

NIELS: Yes, I remember.

JOHN: When did you come to Princeton?

NIELS: Let's see. The first time I came to visit was after the war. I think it was from February to June of 1948, and then I stayed about three months in the early spring of 1950 and again in the late fall of 1954. I think I stayed almost five months that time. Albert and I both did time at the Institute of Advanced Studies, but mostly I just hung out in Albert's office.

ALBERT: Yes, and we seemed to never tire of talking about the nature of reality.

Niels comes over and in a rare display of affection places his hand lightly on Albert's shoulder. Albert turns to Niels and says:

ALBERT: We did have fun didn't we?

NIELS: Yes. And what more can one ask of life?
In his later years Einstein had finally tired of the quantum debate, preferring to spend his remaining years in pursuit of a unified field theory. While he was still able, he walked to his office almost every day and filled his hours in study. Finally, when his legs would no longer carry him, he worked from his bed right up until the day he died on the morning of April 18, 1955. He was 76.

The news of Einstein's death had hit Bohr hard. He wrote of his departed friend: "To the whole of mankind Albert Einstein's death is a great loss and to those of us who had the fortune to enjoy his warm friendship it is a grief that we shall nevermore be able to see his gentle smile and listen to him" (Spangenburg, 98).

Like Albert, Niels was a pacifist and he spent much time in his later years working as an advocate for the peaceful use of nuclear power. Niels died of heart failure at the age of 77, on a Sunday in November of 1962 In Copenhagen. Of Niels, Albert once wrote: "He utters his opinions like one perpetually groping and never like one who believes to be in possession of definite truth" (Spangenburg, 100).

The blackboard scrawlings left by the two men after their death fairly describes their lives. The Confident Einstein, always charging full speed ahead, left equations dealing with his attempt to formulate a unified field theory; whereas, Bohr, still unsure of quantum theory, and perhaps fearful that the great Einstein could turn out to be correct after all, left drawings and equations dealing with objections to quantum theory as presented by Einstein at the 1927and 1930 Solvay conferences.

Rashid awakens early on Sunday morning and, not wanting to disturb Risa, gets up and goes down the hall to see if Albert, Niels, or John have arisen. They too have often used their offices as sleeping quarters. When he gets to John's officer he finds that his name has been removed from the door. Thinking that this is very odd, he continues down the hall only to find that Albert's name and Niels' name have been removed from their doors as well. Rashid now becomes a bit concerned and confused, because he had assumed that the three would be at Princeton for at least a few more days and he wanted the opportunity to say good-bye. Certainly they would not have closed down their offices so abruptly. This just didn't make any sense. Feeling a bit desperate he knocks on Albert's door, not really expecting a response. He paces back and forth, wondering what has happened, and then he sees a janitor coming down the hall pushing a bucket of water on wheels and carrying a mop.

RASHID: Why have the names been removed from these doors? Where are Albert, Niels, and John?

JANITOR: I'm not sure what you're talking about.

RASHID: Are Albert, Niels, and John still here at Princeton?

JANITOR: Son, these offices have been vacant for about three months now. I don't know who these characters Albert, Niels, and John are. You must have the wrong building.

Bewildered, Rashid slowly walks back to his office as he searches for some explanation. When he gets to his office he finds Risa fixing a breakfast of bagels and cream cheese. Rashid, in a very quiet voice speaks slowly and questioningly.

RASHID: Albert... Niels... and John are gone? Albert... Niels... and John are *gone!* This just can't be.

RISA: What are you talking about?

Rashid, in a shaky voice, tries to describe to Risa what has happened and she listens with bemusement.

RISA: Are you dreaming, Rashid? I don't know what you are talking about. Are you thinking about what you read in my manuscript? You were up most of the night reading. Did you think that Albert, Niels, and John were actually here at Princeton? That's really funny. You must have had a really good dream. Would you get the orange juice out the fridge, please?

Rashid is completely miffed. He lies face down on the futon and wonders if Risa might be right. Could all of this have been a dream? Now unsure of anything and doubting his own sanity, he turns onto his back and props up the pillow behind him. He has just one question that could explain everything.

RASHID: Risa, have you thought of a title for your book?

RISA: Ya. I thought I'd call it: *Confronting the Quantum Enigma: Albert, Niels, and John.*

Bibliography

Barbour, Julian. *The End of Time: The Next Revolution in physics. New York:* Oxford University press, 1999

Bell, Eric Temple. *The Magic of Numbers. New York: Dover, 1991.*

Briggs, John, and F. David Peat. *Looking Glass Universe: The Emerging Science of Wholeness.* New, York: Simon & Schuster, 1984.

Brody, Howard. *The Placebo Response: How you can Release the Body's Inner Pharmacy for Better Health.* New York: Harper Collins, 2000

Brooks, Michael. "The Placebo Effect". *New Scientist,* March 19-25, 2005.

Davies, Paul. *About Time: Einstein's Unfinished Revolution.* New York: Simon and Schuster, 1996.

Giorbran, Gevin. *Everything Forever: Learning to See Timelessness.* Seattle: Enchanted Puzzle Publishing, 2007.

Gribbin, John. *In Search of Schrodinger's Cat.* Toronto: Bantam, 1984.

Hebert, Nick. *Quantum Reality: Beyond the New Physics.* New York: Doubleday, 1987.

Isaacson, Walter, *Einstein: His Life and Universe.* New York: Simon and Schuster, 2007.

Joad, C.E.M., *Guide To Philosophy.* New York: Dover, 1957.

Jaynes, Julian, *The Origin of Consciousness in the Break-Down of the Bicameral Mind.* Boston: Houghton Mifflin, 1982.

Lipton, Bruce, *The Biology of Belief: Unleashing the Power of Consciousness, Matter, and Miracles.* Santa Rosa: Mountain of Love/Elite Books, 2005.

Nadeau, Robert and Kafatos, Menas. *The Non-local Universe: The New Physics and Matters of the Mind.* New York: Oxford University Press, 1999.

Nichol, Lee, *The Essential David Bohm. London: Routledge, 2003.*

Nickerson, Raymond S., *Cognition and Chance: The Psychology of Probabilistic Reasoning*. Mahwah, New Jersey: Lawrence Erlbaum Associates, 2004.

Penrose, Roger. *The Emperor's New Mind*. New York: Oxford University Press, 1989.

Schwartz, Jeffrey M. and Begley, Sharon. *The Mind and The Brain: Neuroplasticity and the Power of Mental Force. New York:* Regan Books, Harper Collins, 2002.

Siegfried, Tom. *The Bit and the Pendulum: From Quantum Computing to M Theory—The New Physics of Information*. New York: John Wiley & Sons, 2000.

Spangenburg, Ray and Moser, Diane K. *Niels Bohr: Gentle Genius of Denmark*. New York: Facts on File, 1995.

Turner, Judith A. *"Placebo Effects on Pain": Health Magazine*, April, 1995.

Wade, David. *Crystal & Dragon: The Cosmic Dance of Symmetry & Chaos in Nature, Art, & Consciousness*. Rochester, Vermont: Destiny, 1991.

Wolf, Fred Alan. *Taking the Quantum Leap*. San Francisco: Harper Row, 1981.